INDUCTION COIL
THEORY AND APPLICATIONS

BY

E. TAYLOR JONES, D.Sc.

PROFESSOR OF NATURAL PHILOSOPHY IN
THE UNIVERSITY OF GLASGOW

LONDON
SIR ISAAC PITMAN & SONS, LTD.
1932

SIR ISAAC PITMAN & SONS, Ltd.
PARKER STREET, KINGSWAY, LONDON, W.C.2
THE PITMAN PRESS, BATH
THE RIALTO, COLLINS STREET, MELBOURNE
2 WEST 45TH STREET, NEW YORK

SIR ISAAC PITMAN & SONS (CANADA), Ltd.
70 BOND STREET, TORONTO

PRINTED IN GREAT BRITAIN
AT THE PITMAN PRESS, BATH

PREFACE

THE theory of the action of an induction coil, or that of any other form of oscillation transformer, is essentially a theory of the transient electric currents set flowing at some sudden or very rapid change in the circumstances of one of a pair of coupled circuits. The precise manner of variation of the currents depends upon the method by which they are started, but generally in inductive circuits it takes the form of two superposed oscillations which gradually die away while the system is adjusting itself to its new conditions. In many cases the currents, besides varying with time, are also variable along the wire owing to its distributed capacity—a fact which is too often overlooked, with the consequence that erroneous statements are sometimes made regarding fundamental matters, such as, for example, the law of electromagnetic induction which is discussed in Chapter I.

The book contains a less detailed and more descriptive account of the action of induction coils than that given in the *Theory of the Induction Coil* published eleven years ago. All the essential features of the theory are, however, retained in the present account, and free use has been made of portions of the earlier book where they appeared suitable for the purposes of the present one.

As in the former book, oscillographic records are used largely to illustrate the subject, and many new examples are here collected, including some, in Chapter III, which illustrate the relative merits of coils and transformers as generators of high potentials.

In using an induction coil or other generator for some practical purpose, it is important to understand the nature of the function which it has to perform. One such duty, for which induction coils are in general use at the present time, is that of producing ignition in motor-car engines, and an account of this subject is accordingly given in Chapter VIII, with a discussion of the relative effectiveness in ignition of different types of induction coil spark.

v

The induction coil has recently proved to be a very suitable generator of cathode ray beams for the study of electron diffraction phenomena, and a description of experiments by this method is given in Chapter VI.

At the present time much use is made of the sustained oscillations of coupled circuits, especially those in which the amplitude is kept constant by the action of a triode valve. There is an important difference between such maintained oscillations and the transient vibrations which follow a sudden alteration of the circuit conditions. In the latter, both component vibrations are usually strongly in evidence together, but in the oscillations maintained by a valve only one of the components is usually present, and it is only in very special circumstances that both oscillations can be maintained simultaneously. This question is discussed in Chapter IX, in which the conditions for the maintenance of one component, or the other, or of both together, are explained.

The author wishes to thank the Editors of the *Philosophical Magazine, The Electrician,* and the *Journal of the Röntgen Society,* for their kind permission to use articles and illustrations which have been published in those journals.

Much of the experimental work described in the following pages was carried out in the Physics Laboratory of the University College of North Wales, Bangor, and in the Natural Philosophy Department of the University of Glasgow. To the Council of the College and the University Court of the University of Glasgow the author is much indebted for the facilities which they have given him for engaging in this work.

GLASGOW, 1932.

CONTENTS

vii

CONTENTS

INDUCTION COIL
THEORY AND APPLICATIONS

CHAPTER I

TRANSIENT ELECTRIC CURRENTS IN INDUCTIVE CIRCUITS

THE induction coil is one of a number of arrangements of two coils, called the primary and secondary coils, for producing high potentials by means of transient electric currents. In these arrangements the currents are set going by effecting some sudden change in the connections of the primary circuit, as by breaking the circuit and therefore interrupting the current at a point, or by allowing a condenser to discharge across a spark gap in this circuit. In the induction coil and the high tension magneto the former method, in the Tesla coil and the auto-transformer the latter method, is employed for producing the discontinuity which gives rise to the transient effects.

The action is analogous to that of an air-gun, to which energy is supplied by compressing the spring, and in which this energy is released by pulling the trigger. In the induction coil energy is supplied by the battery in the form of magnetic energy of the primary current. When the current is interrupted at a point this energy is released into the two circuits of the coil. The manner in which the energy becomes distributed among the parts of the system during the transient period after "break" depends greatly upon the capacity of the condenser, which is usually connected across the interrupter, and which should be adjusted so as to produce the most effective results. In the Tesla coil the energy is supplied as electrostatic energy of a charged condenser in the primary circuit, and the spark effects the sudden release of this energy into the circuits.

In all these arrangements the secondary coils, being made of great lengths of wire, have considerable electrostatic capacity without the addition of condensers. All have, therefore, certain features in common, viz. in each of them the primary

1

and secondary circuits both contain self inductance and capacity, and the primary and secondary coils act upon each other by their mutual magnetic induction, that is, they are magnetically "coupled."

Each of the systems we are considering consists, therefore, of a pair of coupled oscillatory circuits, and the transient effects produced in them usually consist of two superposed electrical oscillations, differing in frequency and amplitude, which combine if circumstances are favourable to produce the required high potentials. Such systems are known as "oscillation transformers," and the theory of their action is an application of the general theory of electrical oscillations in coupled circuits. This theory covers most of the cases, even those in which the coils are wound on iron cores, provided these cores do not form closed magnetic circuits. If the cores are closed, the inductances cannot be treated as nearly constant, and the dissipative forces, which cause excessive damping of the oscillations, are not subject to simple laws, so that the theory requires modification.

The action of oscillation transformers is very different from that of the ordinary transformer which, as generally used, is supplied in a continuous manner with alternating current from an external source, and in which no interruption or other discontinuity of action occurs. It is true that oscillation transformers can be excited by supplying them with alternating current from an A.C. generator or a triode oscillator circuit, and certain interesting effects are obtained in this way (see Chapter IX), but their usual action consists in the play of their free or natural oscillations which die away more or less rapidly in accordance with the damping forces of the system. This free movement of the electricity in the system, which is supplied with energy and then left to itself, is one of the principal features of the action of induction coils and the other forms of oscillation transformer which have been mentioned.

Electrical Oscillations. The theory of electrical oscillations in a single circuit was first given in 1853 by William Thomson* (afterwards Lord Kelvin), who showed that if the plates of

---

* *Phil. Mag.*, June, 1853. The theory was first established for a system consisting of a single conductor connected by a wire with the earth.

a condenser of capacity C are initially charged and are connected by a wire of resistance R and self inductance L, the electricity will oscillate to and fro in the circuit with a frequency given by the expression

$$n = \frac{1}{2\pi} \sqrt{\frac{1}{LC} - \frac{R^2}{4L^2}} \qquad . \qquad . \qquad . \qquad . \qquad . \qquad (1)$$

which reduces, if the term $R^2/4L^2$ is small in comparison with $1/LC$, as it is in many cases, to the well-known simpler expression

$$n = 1/2\pi\sqrt{LC} \qquad . \qquad . \qquad . \qquad . \qquad . \qquad . \qquad (2)$$

It was also shown by Kelvin that the oscillations die away at a rate depending upon the value of $R/2L$, which is called the damping factor of the oscillations. The expression for the potential difference V of the plates of the condenser at any time t after the oscillations begin is of the form

$$V = Ae^{-Rt/2L} \sin(2\pi nt - 0) \qquad . \qquad . \qquad . \qquad . \qquad (3)$$

the constants A and 0 depending upon the initial conditions, i.e. upon the manner in which the oscillations are started. The current i at any time can be found from the above expression for V by the relation

$$i = C\frac{dV}{dt} \qquad . \qquad . \qquad . \qquad . \qquad . \qquad . \qquad . \qquad (4)$$

which expresses that the current in the coil is equal to the rate at which the charge of the condenser is increasing.

To apply these results to a particular problem, let us suppose that the circuit is provided with an interrupter I (Fig. 1), which has a condenser of capacity C connected to its terminals. Initially, the circuit is closed at I and a current i_o is maintained in it by a battery of E.M.F. E. At a certain moment $t = o$, from which time is reckoned, contact is broken at I so that the current is interrupted at that point. The current in the coil, however, does not experience any sudden change, but continues to flow now into the condenser, which becomes charged as the current in the coil gradually diminishes to zero. When the condenser has reached its maximum potential, it begins to discharge back through the coil, the current in which is now reversed. In this way, the current is caused to oscillate n the circuit, its amplitude gradually diminishing owing to the dissipation of energy arising from the resistance of the circuit.

The equations for the circuit are

$$L\frac{di}{dt} + Ri + V = E \quad . \qquad . \qquad . \qquad . \qquad . \qquad . \qquad (5)$$

$$i = C\frac{dV}{dt}$$

and the initial conditions of the problem are

$$i = i_o, \ V = o, \text{ when } t = o \quad . \qquad . \qquad . \qquad . \qquad . \qquad (6)$$

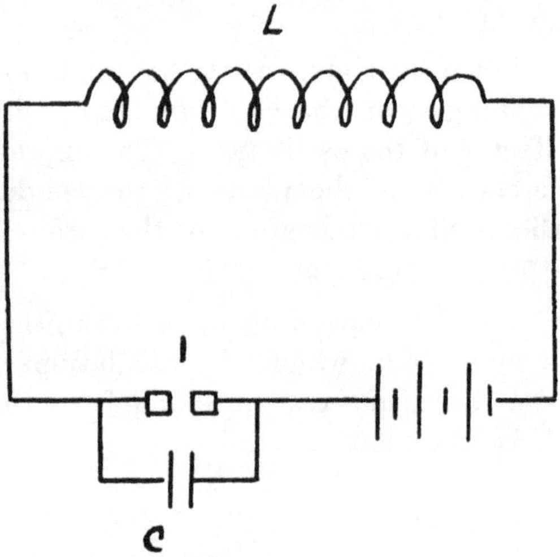

Fig. 1. Oscillatory Circuit with Battery and Interrupter

It can be shown that the complete solution, giving the potential difference of the plates of the condenser at time t, is

$$V = E + A\,e^{-\,Rt/2L}\sin\,(2\pi nt - \theta) \quad . \qquad . \qquad . \qquad (7)$$

where

$$A^2 = E^2 + \frac{1}{4\pi^2 n^2}\left(\frac{i_o}{C} - \frac{ER}{2L}\right)^2, \quad . \qquad . \qquad . \qquad (8)$$

$$\tan\theta = 2\pi n\bigg/\left(\frac{1}{RC} - \frac{R}{2L}\right)$$

and the frequency n is given by equation (1). The transient wave of potential, i.e. the oscillation, is represented by the second term in (7). This dies away in accordance with the exponential factor, and in the final state the condenser is charged to the constant potential E, the electromotive force of the battery.

If the resistance of the circuit is very small the E.M.F. E should also be small, otherwise the steady current i_o would be extremely large. Neglecting both E and R therefore in the above expressions we find $\theta = O$, and

$$V = \frac{i_o}{2\pi nC} \sin 2\pi nt, \qquad . \qquad . \qquad . \qquad . \qquad . \qquad (10)$$

the frequency n being now given by (2).

As another problem of this kind we may consider the excitation of oscillations by a spark produced between the electrodes

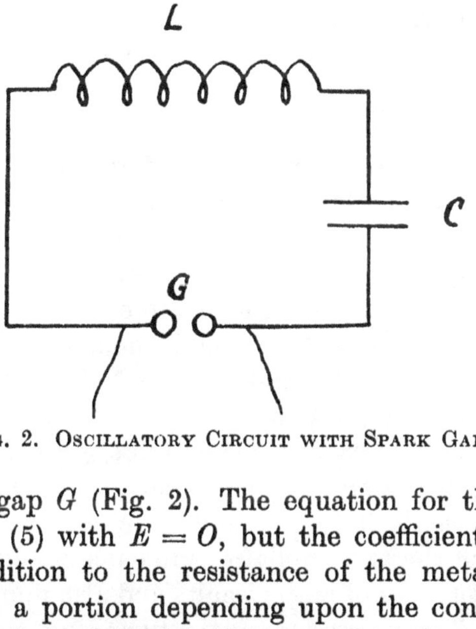

FIG. 2. OSCILLATORY CIRCUIT WITH SPARK GAP

of the spark gap G (Fig. 2). The equation for the circuit is now equation (5) with $E = O$, but the coefficient R now includes, in addition to the resistance of the metallic portion of the circuit, a portion depending upon the conductivity of the air gap G through which the current in the spark is conveyed. The initial conditions are

$$V = V_o, \text{ the sparking potential of the gap,} \left.\begin{array}{c} \\ \\ \end{array}\right\} \qquad . \qquad . \quad (11)$$
$$i = o, \text{ when } t = o$$

The solution is found to be

$$V = V_o\, e^{-Rt/2L}\left(\cos 2\pi nt + \frac{R}{4\pi nL} \sin 2\pi nt\right) \qquad . \qquad . \quad (12)$$

The transient effect on the potential in this case is an oscillation of frequency n given by (1), and of initial amplitude equal to the sparking potential of the gap. This method, in which the oscillation is started by a spark, is one of the most convenient for producing high frequency oscillations. If the self inductance L is 0·005 henry, and the capacity C is 0·002 microfarad, the frequency is about 50,000 oscillations per sec. It may be remarked that if in such a case the spark is produced by means of an induction coil connected to the spark gap terminals, the presence of the coil has practically no effect on the high frequency current. The self inductance of the secondary of the coil is so great that no appreciable fraction of the high frequency current passes through it.

From the solution for V in any such problem the value of the current i at any time during the oscillation can be found by the relation $i = C\dfrac{dV}{dt}$.

Electrical Oscillations in Coupled Circuits. Turning now to the question of the electrical oscillations in a pair of coupled circuits, each containing self inductance and capacity, we find that since any variation of the current in one circuit induces an E.M.F. in the other, the current in each circuit will oscillate with two frequencies which will in general differ from each other. Further, since the inductive E.M.F. in each circuit depends upon mutual as well as upon self inductance, the frequencies of the two oscillations will not generally be the same as those which the circuits would have if they were removed from each other's inductive influence. In other words, the frequencies of the two oscillations in each circuit depend upon the self inductances of both circuits and also upon their mutual inductance. The expression for the two frequencies, in the case when the resistances are negligible, was first given by Oberbeck.* If the self inductances and capacities of the primary and secondary circuits are distinguished by suffixes, the mutual inductance is denoted by M, and the ratio M^2/L_1L_2, known as the "coupling" of the circuits, is written k^2, the expression is

$$8\pi^2 n^2 = \frac{1}{1-k^2}\left[\frac{1}{L_1C_1} + \frac{1}{L_2C_2} \pm \sqrt{\left\{\left(\frac{1}{L_1C_1} - \frac{1}{L_2C_2}\right)^2 \right.}\right.$$
$$\left.\left. + \frac{4k^2}{L_1C_1L_2C_2}\right\}\right] \qquad . \qquad . \qquad . \qquad . \qquad (13)$$

* *Wied. Ann.*, **55**, p. 623, 1895.

We will denote the greater of the two frequencies by n_2, the smaller by n_1. It is convenient in many calculations to represent the ratio L_1C_1/L_2C_2 by a single symbol. Calling this u, we find for the ratio of the squares of the two frequencies

$$\frac{n_2^2}{n_1^2} = \frac{1 + u + \sqrt{\{(1 - u)^2 + 4k^2u\}}}{1 + u - \sqrt{\{(1 - u)^2 + 4k^2u\}}} \qquad . \qquad . \qquad (14)$$

It is easy to verify that the frequency ratio n_2/n_1 is smallest when $u = 1$, its value then being equal to $\sqrt{\dfrac{1 + k}{1 - k}}$.

It appears from these expressions that the two frequencies of oscillation of a coupled system, in which the resistances are small, cannot be equal, and that the ratio of the frequencies n_2/n_1 has a minimum value determined by the coupling. If the coupling k^2 is 0·36, the smallest frequency ratio is 2, if k^2 is 0·64 the smallest ratio is 3, if k^2 is 0·779 the least ratio is 4, and if k^2 is 0·852 the minimum is 5. In a coupled system the frequency ratio can be varied over wide limits by changing one or other of the capacities C_1 and C_2, but the ratio cannot be less than the value $\sqrt{(1 + k)/(1 - k)}$ which it has when $u = 1$, i.e. when $L_1C_1 = L_2C_2$. The relation $u = 1$ is the condition that the two circuits should have the same frequency of oscillation when they are separated from one another. We conclude that two circuits which have the same *single* frequency when separated have the same *two* frequencies when brought near each other so as to have a certain coupling, the ratio of the frequencies being the minimum for this coupling. One of the frequencies is greater, the other less, than the frequency of the circuits when separated. If the frequencies of the separated circuits are not equal, that is if L_1C_1 is not equal to L_2C_2, then one of the frequencies of the coupled system is greater, the other less, than *both* of the separate circuit frequencies. In other words, in any magnetically coupled system in which the influence of the resistances on the frequencies is small, neither of the two frequencies lies between the values $1/2\pi\sqrt{L_1C_1}$ and $1/2\pi\sqrt{L_2C_2}$; one frequency is greater, the other less, than both of these values.

There are many mechanical systems which possess similar oscillatory properties. A pair of simple pendulums, one suspended from the bob of the other, oscillate with two frequencies, which differ from those which the pendulums have when suspended independently. A weight suspended by a spiral spring, with another spring and weight attached to it, is another such system, each of the two weights making vertical oscillations of two frequencies. Another example is found in a pair of pulleys capable of turning on a fixed horizontal axle, the rotation of each pulley being controlled by a spring. An endless cord passes over both pulleys and carries in each bight a loose pulley supporting a weight. It was shown by the late Lord Rayleigh* that the mutual action of the two pulleys of this system is very closely analogous to that of two magnetically coupled circuits. All such systems consisting of two coupled oscillators and, therefore, having two degrees of freedom, whether they are mechanical or electrical systems, and whatever be the nature of the coupling between their parts, have frequency relations similar to those just described.

Circuit Equations. In forming the circuit equations for an oscillation transformer, we must bear in mind two circumstances which have the effect of rather complicating the problem. The first is the fact, already mentioned, that the terminals of the secondary coil are not usually connected with a condenser, and that the capacity of the secondary circuit is mainly if not entirely the distributed capacity of the wire forming the secondary coil. In many cases, both terminals of the secondary coil are insulated, and in these circumstances the current in this coil during the oscillations is greatest in the central winding and zero at the ends. The distribution of the current in the coil is analogous to that of the velocity in a stretched vibrating string sounding its fundamental tone, the motion being greatest at the centre of the string and zero at the ends. Employing a notation first used by Drude† we shall represent by i_2 the current in the central winding of the secondary coil, and by V_2 the potential difference of the terminals of this coil. The secondary capacity C_2 is a capacity determined by the

---

* *Phil. Mag.*, XXX, p. 30 (1890).
† *Ann. d. Physik*, 13, p. 512 (1904).

size and form of the secondary coil and defined by the equation—

$$i_2 = C_2 \frac{dV_2}{dt}.$$

It may also be defined as the charge distributed over one half of the secondary coil and the bodies connected with its terminal divided by the difference of potential of the terminals.

The unequal distribution of current in the secondary wire has also other effects. The self induction of the secondary coil during the oscillations is clearly smaller than it would be if the current had the value i_2 in all parts of the coil. Again following Drude, we define the self inductance L_2 of the secondary coil as the magnetic flux through this coil due to the current in it divided by the current i_2 in the central winding. The electromotive force of self induction in the secondary coil is, therefore, represented by $L_2 \dfrac{di_2}{dt}$.

Further, the inductive effect of the secondary current on the primary during the oscillations is smaller than it would be if the current were uniformly distributed and equal to i_2. The inductance of the secondary on the primary, L_{12}, is defined as the magnetic flux through the primary coil due to the secondary current divided by the value of this current in the central winding. The induced E.M.F. in the primary due to variation of the secondary current is, therefore, represented by $L_{12} \dfrac{di_2}{dt}$. In the primary circuit, which is associated with a condenser of large capacity, the current may be regarded as uniformly distributed along the wire. The self inductance L_1 of the primary circuit has, therefore, the usual meaning, and the inductance of the primary on the secondary, L_{21}, is equal to the mutual inductance of the coils as usually defined.

The coupling k^2 of the two circuits is defined as the ratio $L_{21}L_{12}/L_1L_2$. It is found by experiment that the coupling of the two circuits of an induction coil is appreciably smaller when the coil is used without a secondary condenser than it is when the secondary terminals are connected to a condenser of considerable capacity, in which case the secondary current can be regarded as uniformly distributed.

If one end of the secondary coil is earthed, i_2 should be taken as the current at the earthed end. The same symbols may be used, but the values of the capacity and induction coefficients will differ from those which they have in the symmetrical arrangement in which both terminals are insulated.

The other complication arises from the fact that electrical oscillations in a coil are subject to decay due to other causes than the ohmic resistance of the coil. These causes are : (1) Core losses (eddy currents and hysteresis) in systems having iron cores ; (2) leakage between the plates of condensers and due to brush discharge from high potential terminals. In a well-constructed induction coil, the losses due to these causes are considerably greater than those due to the ohmic resistances of the wires. Nevertheless, it is found that, provided the losses are not exceedingly great, they can be taken into account sufficiently well by the inclusion in the equation of factors representing what are called the effective resistances of the circuits. The effective resistance of an oscillatory circuit usually depends upon the frequency, and in high frequency circuits the ohmic resistance itself is considerably increased by the "skin effect." It is to be understood that the coefficients R_1, R_2 in the equations represent, not the resistances of the two circuits for steady currents, but their effective resistances, that is, factors which determine the mean values of the energy loss due to all causes. The inductances are also usually not strictly constant, especially in iron core systems, and the symbols representing them are to be understood as mean values suitable for the circumstances of the problem under consideration.

With these definitions the circuit equations may be written

$$L_1 \frac{di_1}{dt} + L_{12} \frac{di_2}{dt} + R_1 i_1 + V_1 = 0 \qquad . \qquad . \qquad . \quad (15)$$

$$L_2 \frac{di_2}{dt} + L_{21} \frac{di_1}{dt} + R_2 i_2 + V_2 = 0 \qquad . \qquad . \qquad . \quad (16)$$

$$i_1 = C_1 \frac{dV_1}{dt} \quad , \quad i_2 = C_2 \frac{dV_2}{dt} \qquad . \qquad . \qquad . \quad (17)$$

where V_1 represents the potential difference of the plates of the primary condenser, or more exactly the excess of this quantity over the E.M.F. of the battery.

If i_1, i_2 are eliminated, the equations become

$$L_1 C_1 \frac{d^2 V_1}{dt^2} + L_{12} C_2 \frac{d^2 V_2}{dt^2} + R_1 C_1 \frac{dV_1}{dt} + V_1 = 0 \qquad . \quad (18)$$

$$L_2 C_2 \frac{d^2 V_2}{dt^2} + L_{21} C_1 \frac{d^2 V_1}{dt^2} + R_2 C_2 \frac{dV_2}{dt} + V_2 = 0 \qquad . \quad (19)$$

With suitable modifications in certain cases, and with the appropriate initial conditions, these equations allow the primary or the secondary potential at any moment during the flow of the transient currents to be calculated for any form of oscillation transformer.

For the present we shall confine our attention to the problem of the induction coil, worked by an interrupter, the initial conditions for the oscillations set up at break (i.e. at $t = 0$) being in this case—

$$\left. \begin{array}{l} i_1 = C_1 \dfrac{dV_1}{dt} = i_0, \\[2mm] i_2 = C_2 \dfrac{dV_2}{dt} = 0, \\[2mm] V_1 = -E, \\[2mm] V_2 = 0 \end{array} \right\} \qquad . \qquad . \qquad . \qquad . \qquad . \qquad (20)$$

In general, if the resistances R_1, R_2 are not restricted in value, the equations do not admit of simple algebraic solutions, but if the resistances are small, approximate solutions can be found. In this case, the influence of the resistances on the frequencies can be neglected, since it depends on the squares of the resistances, and the frequencies are then given by the expression (13). The damping factors of the oscillations can be determined from approximate expressions given by Drude, which will be referred to later. (See Chapter III.)

Certain problems on the action of induction coils, such as that of determining the conditions in which a coil will give the greatest possible secondary potential, can be solved without reference to the resistances, since they depend mainly upon the frequency relations of the system. We shall, therefore, first consider the solution of equations (18) and (19) for the case in which the resistance terms are altogether neglected. Omitting the terms containing R_1 and R_2 we find the solutions to be—

$$V_2 = \frac{2\pi L_{21}\, i_0\, n_1 n_2}{n_2{}^2 - n_1{}^2}\left(n_2 \sin 2\pi n_1 t - n_1 \sin 2\pi n_2 t\right) \qquad . \quad (21)$$

$$V_1 = \frac{2\pi i_0\, n_1 n_2{}^2}{C_1(n_2{}^2 - n_1{}^2)}\left(L_1 C_1 - \frac{1}{4\pi^2 n_2{}^2}\right)\sin 2\pi n_1 t$$
$$-\frac{2\pi i_0\, n_1{}^2 n_2}{C_1(n_2{}^2 - n_1{}^2)}\left(L_1 C_1 - \frac{1}{4\pi^2 n_1{}^2}\right)\sin 2\pi n_2 t \quad . \qquad . \quad (22)$$

The wave of potential in the secondary circuit after "break" thus consists of two oscillatory components the amplitudes of which are inversely proportional to the frequencies. In the primary circuit the potential difference of the plates of the condenser also oscillates with two frequencies, but the relation between the amplitudes is less simple than that found in the secondary wave.

From these solutions we find the expressions for the currents

$$i_1 = C_1\frac{dV_1}{dt}$$
$$= \frac{4\pi^2 i_0\, n_1{}^2 n_2{}^2}{n_2{}^2 - n_1{}^2}\left(L_1 C_1 - \frac{1}{4\pi^2 n_2{}^2}\right)\cos 2\pi n_1 t$$
$$-\frac{4\pi^2 i_0\, n_1{}^2 n_2{}^2}{n_2{}^2 - n_1{}^2}\left(L_1 C_1 - \frac{1}{4\pi^2 n_1{}^2}\right)\cos 2\pi n_2 t \quad . \qquad . \quad (23)$$

and

$$i_2 = C_2\frac{dV_2}{dt}$$
$$= \frac{4\pi^2\, L_{21} C_2\, i_0\, n_1{}^2 n_2{}^2}{n_2{}^2 - n_1{}^2}\left(\cos 2\pi n_1 t - \cos 2\pi n_2 t\right) \qquad . \quad (24)$$

The two components of the oscillatory current in the secondary coil have, therefore, equal amplitudes. Since we are at present supposing that no discharge passes between the secondary terminals, the whole of the electricity flowing in the current i_2 is accumulated as electrostatic charge in the secondary circuit. The amount of this charge, that is, the charge on the

positive portion of the secondary coil and any insulated conductors that may be connected with its terminal, at any
moment during the oscillations is represented by the expression
(21) multiplied by C_2. The action of an induction coil when
producing discharge will be considered in Chapter V.

Since the primary wire is wound closely upon the iron core,
the magnetic flux in the core multiplied by the number of
turns in the primary coil is $L_1 i_1 + L_{12} i_2$, which is found to be
equal to—

$$\frac{n_2^2 \, i_0}{C_1(n_2^2 - n_1^2)} \left(L_1 C_1 - \frac{1}{4\pi^2 n_2^2} \right) \cos 2\pi n_1 t$$

$$- \frac{n_1^2 \, i_0}{C_1(n_2^2 - n_1^2)} \left(L_1 C_1 - \frac{1}{4\pi^2 \, n_1^2} \right) \cos 2\pi n_2 t. \qquad . \qquad . \quad (25)$$

All these solutions, (21) to (25), represent undamped oscillations, and they, therefore, do not correctly represent the
oscillations of an actual induction coil which are subject to
considerable damping. The above expressions along with (13)
do, however, give with considerable accuracy the initial values
of the amplitudes, as well as the phases and frequencies, of the
oscillations which take place in a well-constructed induction
coil after "break," and they serve sufficiently well for the determination of the effect of varying one or other of the inductances or capacities of the system. Before entering upon the
further discussion of these solutions we shall consider certain
matters relating to the construction and use of induction coils
and to the historical development of views as to the nature
of their action.

Construction of Induction Coils. The first experiment in
which a high potential effect was obtained in a secondary coil
by making or breaking a primary circuit was made by Faraday *
in 1831. The apparatus used by Faraday consisted of an iron
ring on which were wound two coils of wire, insulated from
each other and from the ring, one of the coils being connected
with a battery. By making or breaking the connection between the battery and this coil, Faraday produced short sparks
between charcoal terminals connected with the other coil. He
observed that the sparks were produced more readily when

* *Phil. Trans.*, p 132 (1832).

contact was made than when it was broken. We shall see
later (Chapter IV) that the circuits of any induction coil can
be adjusted so that the secondary potential at "make" is
greater than that at "break." Usually, however, it is desired
to produce a much higher potential at break than at make.

Following on Faraday's discovery, other experimenters soon

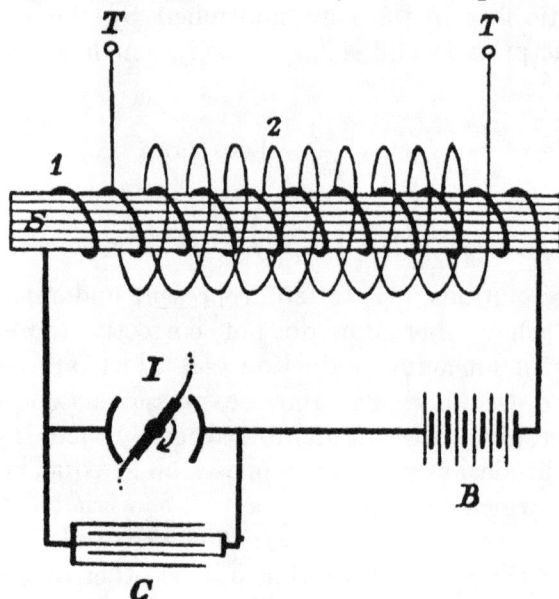

FIG. 3. CIRCUITS OF INDUCTION COIL

S = Core.   I = Interrupter.
    1, primary coil;   C = Condenser.
    2, secondary coil.   T T = Secondary terminals.
B = Battery.

began to develop the induction coil, one of the first improve-
ments being the substitution of a straight core of iron wire
for the iron ring. The advantage gained in this way is due
to the great diminution of the eddy current losses in the core,
and therefore of the damping of the oscillations.

A further great improvement was effected in 1853 by Fizeau,
who first used a condenser in conjunction with the interrupter.
We shall see that the condenser not only diminishes the tendency to arcing at the interrupter, but also, by suitable choice
of its capacity, allows the circuits to be adjusted so as to produce the most effective results.

Fig. 3 shows a diagram of the circuits of an induction coil.

In modern coils the core S is usually straight, and is built up of insulated wire or sheet of silicon steel. This substance has the important magnetic properties of high permeability and low hysteresis, and its high specific resistance tends further to diminish the loss due to eddy currents. Instead of the straight core, some makers prefer to construct their coils with an iron core bent so as to form a nearly closed magnetic circuit. A step in the same direction was suggested by Dessauer,* who recommended the use of wide flanges of iron attached to the ends of a straight core. The effect of partly closing the iron circuit is to increase all the inductances and also to increase the coupling. Such variations are fully taken into account in the theory, and will be considered in Chapter II. The partial closing of the iron circuit also tends to increase the core losses and therefore the damping of the oscillations, and this always diminishes the maximum secondary potential attainable with a coil. So long as the iron circuit is not too completely closed, however, this latter effect is not sufficient to interfere seriously with the working of the coil. Completely closed iron cores are never used with induction coils, though this is the usual mode of construction in the high tension transformer, an instrument now much used (with alternating current and rectifying valves) for X-ray production. In spite of the very rapid damping of the oscillations of a high tension transformer, however, one of these instruments, if worked by a battery and interrupter, gives a much higher secondary potential than it would if supplied with alternating current of peak value equal to that supplied by the battery. (See Chapter III.)

On the core, and well insulated from it, is wound the primary coil (1, Fig. 3) of fairly thick copper wire.† This coil is sometimes tapped at two or three points so that, by means of a commutator, its sections can be connected with one another all in series or all in parallel or in other ways. The question of the best method of connection of the sections, with regard to the production of high secondary potential at break, and low potential at make, is considered in Chapter IV.

---

* *Phys. Zeits.*, 22, p. 425 (1921).
† In ignition coils the primary wire is usually wound outside the secondary. See a paper by E. A. Watson on Coil Ignition Systems, *Journ. I.E.E.*, 1932.

Over the primary coil is placed the primary tube (not indicated in Fig. 3), usually of ebonite or micanite about 1 cm. thick.

Outside the tube is placed the secondary coil (2, Fig. 3), consisting of a very large number of turns of thin copper wire, usually wound in flat sections each 2 or 3 mm. thick. These are generally wound separately and then assembled, with an annular disc of thick paper or other insulating material separating each section from the adjacent ones, the ends of the wire being soldered together so that it forms one continuous winding in the same direction. In some coils the sections are connected together alternately at the inner and outer edge, in others the outer turn of one section is joined to the inner of the next. When the required number of sections are thus put together the whole is thoroughly impregnated with wax, a process which is, in order to remove air-holes, carried out under reduced pressure. The width of the insulating disc, measured radially from the inner to the outer edge, should be considerably greater than that of the wire sections so that there is some distance between the innermost turns of the secondary wire and the primary tube. This is in order to diminish the tendency to sparking near the tube from one section to another, but it also has the effect of reducing the electrostatic capacity of the secondary coil—a quantity which depends upon the interval between the primary and secondary wires—and also of reducing the coefficient of coupling of the two coils. As we shall see later, these two quantities are of great importance in connection with the theory of the action of an induction coil. The separating discs should also extend at the outer edge beyond the circumference of the sections themselves. The ratio of the numbers of turns in the secondary and primary varies greatly in coils by different makers. In many cases it is about 100, but it is as great as 500 in others.

The condenser (C, Fig. 3) consists of a number of sheets of tinfoil separated by a dielectric of waxed or varnished paper—sometimes mica is used—the alternate sheets of foil being connected to one side or the other of the interrupter. The most important requirement in the condenser is that the strength of the dielectric should be sufficient to enable it to withstand the greatest electric strain to which it is liable to be submitted.

When a large coil is in use the potential difference at the plates of the condenser may rise to some thousands of volts —in one case, R. S. Wright observed a $\frac{1}{2}$ in. spark between the ends of the primary winding*—and puncturing of the condenser dielectric is a frequent cause of failure in the performance of coils. Between two consecutive sheets of tinfoil there should be at least two thicknesses of paper, to reduce the probability of any small perforations extending right across the dielectric, and great care should be taken in the manufacture of condensers to exclude moisture and impurities. When a coil is to be used for different purposes or when the connection of the primary layers is to be varied, it is an advantage to be able to vary the capacity of the condenser. For this purpose subdivided condensers are obtainable, in which the various sections can be employed singly or in groups.

The above description of constructional details applies to the usual type of coil which is a familiar piece of apparatus in every physics laboratory, but there is considerable variety both in the manner of construction and in the mounting of coils. In the usual type, the coil is mounted with its axis horizontal, but some large coils are mounted vertically, sometimes, for better insulation, immersed in oil, a method of insulation which is generally adopted in high tension transformers. There are also differences in the manner of winding the secondary coil. The most usual winding is in flat sections three or four wires thick, but some coils have only two secondary sections, others a very large number of sections, each only one wire thick.

Interrupters. In the oldest type of contact breaker, both contact pieces were of solid metal, and this type, with platinum or tungsten contacts, is still in use with small laboratory coils and in coils used for ignition. But, with large coils and heavy currents it is quite unsuitable owing to the effects of the spark which is liable to occur at the point of interruption and which rapidly disintegrates the contact pieces. In such cases mercury interrupters are much more satisfactory and are now in general use.

The older form of mercury break consisted of a rod or dipper

* *Journ. Rönt. Soc.*, IX, 35, p. 4 (1913).

of metal, which was caused to oscillate in a vertical or inclined direction and to make contact in each oscillation with the surface of some mercury under oil or alcohol in a glass vessel. In some forms of the arrangement the movement was maintained by the magnetism of the core, in others it was produced by an independent electric motor. The latter method has the advantage that the movement of the dipper is quite independent of the current flowing in the primary coil.

The more recent forms of mercury interrupter, however, are rotary in action, among the most popular types being those known as the "centrifugal" and the "jet," though the working of both depends upon centrifugal action. In the former an iron vessel containing mercury is set into rapid rotation about a vertical axis by an electric motor, the mercury becoming raised at the sides in accordance with a well-known mechanical principle. In one form of the instrument contact is made during a part of the revolution with this raised belt of mercury by an horizontal metal rod, also rotating, but about an axis parallel to and at a short distance from that of the vessel. An adjustment of the distance between the axes allows the duration of contact to be varied. In these interrupters the mercury is generally covered with a layer of petroleum or other insulating liquid. It is claimed for this type of break that, owing to the rotation, the mercury in the neighbourhood of the place of contact is kept much cleaner than it is in interrupters of the ordinary dipper variety.

In one type of jet interrupter (*I*, Fig. 3) mercury is pumped by centrifugal action up two rotating tubes, inclined upwards and outwards from near the bottom of the containing vessel, and projected horizontally in the form of two revolving jets which impinge on a set of fixed metallic blades of triangular form. The duration of contact can be varied by raising or lowering the blades, thus allowing a narrower or a wider part of their surface to be swept by the jets. The frequency of the interruptions depends upon the rate of rotation of the jets and upon the number of blades. As many as 150 breaks per second may be effected with this interrupter, which is, owing to the very short time of contact, suitable for use on a high voltage circuit. In order to ensure good interruptions it is essential that the mercury surface should be clean, and for this reason coal gas is generally used as the dielectric in these interrupters. Owing to the large number of discharges per second, rapid jet interrupters give very steady illumination in X-ray screen work, but since they require a high voltage supply to enable the primary current to rise sufficiently during the very short time of contact, they are apt to give rise to considerable negative potential and reverse current at "make." (See Chapter IV.)

In some experiments it is desirable to use single discharges produced by the interruption of a strong current in the primary circuit. In using a mercury dipper interrupter for such a purpose, Mr. J. Meiklejohn, a research student in the Natural Philosophy department of the University of Glasgow, observed that the mercury surface is much more easily kept clean if it is convex than if it is plane. The particles of black deposit which tend to collect on the surface are unstable at the summit and easily move down to the side. A simple form of interrupter in which this idea is embodied is illustrated in Fig. 4. It consists of a cylindrical iron vessel, about $1\frac{1}{4}$ in. internal diameter, having

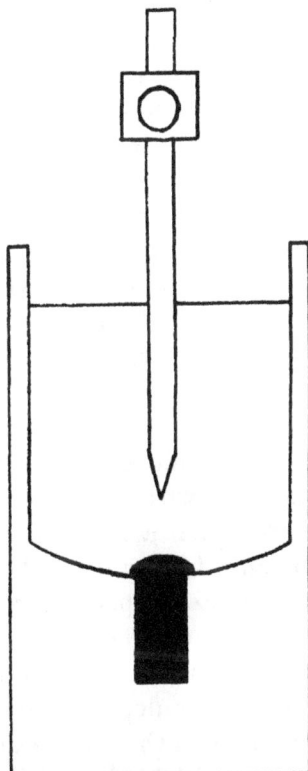

FIG. 4. MERCURY DIPPER INTERRUPTER

a concave bed with a well bored in it at the centre to contain the mercury. The convex surface of the mercury stands above the level of the centre of the bed and is well covered with paraffin oil. Contact is made by an amalgamated copper rod which is attached to the end of a lever and, when not in use, is held by a spring up out of the mercury. If the mercury surface is wiped with a strip of cardboard before the experiment a good break, for currents of 25 amp. or more, is ensured.

If the experiment requires the interruption of a much stronger current the method described by C. Déguisne* may be employed, in which the current is increased until a fuse is blown out in the primary circuit. In this method the primary current may rise to 250 amp., and produce at break a momentary current of 0·4 amp. through an X-ray tube connected with the secondary terminals.

Electrolytic Interrupters. The electrolytic interrupter was first used by Wehnelt in 1899. In its simplest form the Wehnelt interrupter consists of two electrodes, one small and of platinum, the other of lead and large in area, both immersed in a vessel containing dilute sulphuric acid. The platinum electrode may be a piece of wire sealed into the lower end of a glass tube containing mercury to lead in the current. The interrupter is connected, in series with the primary coil, to a battery or other source of high E.M.F., the platinum wire to the positive pole. The current then becomes very rapidly made and broken, the frequency depending mainly upon the E.M.F., the self-inductance of the coil, and the area of the platinum electrode. Usually several hundreds per second, the frequency may be as high as 2,000 with this interrupter, which is chiefly used in X-ray work when a large number of discharges in a short interval of time is required. The least voltage required to work the interrupter depends largely upon the temperature of the electrolyte, and may be as low as 20 volts or so if the liquid is hot. There is also an upper limit to the voltage required; both limits depend not only on the temperature of the liquid but also on the self-inductance of the primary circuit. In some forms of Wehnelt interrupter the platinum electrode, supported by a metal rod, projects through a small opening at the lower end of a porcelain tube and is adjustable; in other forms two or more anodes are provided.

The precise mode of action of the Wehnelt interrupter does not appear to have been fully ascertained. One view of the matter, probably the one which is most generally held, is that the interruption of the current is due to the rapid formation of a layer of gas (oxygen and water-vapour) round the platinum electrode, and that the primary spark which shortly afterwards pierces the layer is responsible for the removal of this barrier

* *Phys. Zeitsch.*, 15, p. 630 (1914).

and the re-establishment of contact between the liquid and the platinum. To judge by an oscillogram of the primary current given by Salomonson* it appears that in some cases contact follows very quickly after break, not infrequently before the current has had time to fall to zero. In this respect the action of the Wehnelt interrupter is very different from that of other interrupters, in which, as a rule, the current not only falls to zero but also performs a number of oscillations about the zero value before the next contact is made.

It is held by some that, in addition to acting as a break, the Wehnelt interrupter also performs the function of a condenser, but about this there is difference of opinion. The condenser usually connected in parallel with interrupters of other kinds is seldom employed with the Wehnelt, which is said to work quite as well, if not better, without the condenser. In some cases examined by the writer, the addition of a condenser (with a given E.M.F. and primary resistance, and with secondary sparks passing) was found to have the effects of (1) lowering the frequency of the interruptions, (2) increasing the mean secondary current. The quantity of electricity discharged across the spark-gap in each cycle must therefore have been greater with than without the condenser. The whole question of the action of electrolytic interrupters appears, however, to require further detailed investigation.

Another form of electrolytic break is that invented independently by Simon and Caldwell in 1899, in which both electrodes are of lead and of large area, but they are placed in separate vessels containing acid and communicating with each other only through a small aperture. In this symmetrical arrangement the interruptions occur at the aperture, where the cross-section of the conducting liquid is small.

Electrolytic interrupters are usually very inefficient in working, a large part of the total energy expended, over 80 per cent in some cases, appearing as heat in the electrolyte. An improved form in which this defect is greatly diminished has been described by F. H. Newman.† In this form the electrolyte

---

* *Jour. Rönt. Soc.*, VII, 27, Fig. 20, p. 12. See also Armagnat, *La Bobine d'induction*, p. 66, and Fig. 37 in this book.
† *Proc. Roy. Soc.* A, 99, p. 324 (1921).

is a saturated solution of ammonium phosphate contained in an aluminium vessel which acts as cathode, the anode consisting as usual of a platinum wire. This interrupter is said to work at a much smaller current than the Wehnelt, and without disintegration of the platinum wire, and to give a very steady series of high potential waves in the secondary coil.

Spark Gaps. A very useful accessory in the use of high tension apparatus is a good adjustable spark gap for measuring the maximum, or peak, value of the secondary potential. The gap should have spherical electrodes, should be mounted on a rigid ebonite stand, and when in use should not be very near other conductors. Electrode diameters of 2 cm. are quite suitable for spark lengths up to 2 or 3 cm. ; for higher potentials larger spheres are sometimes used.

The curve in Fig. 5 shows the sparking potentials in kilovolts for various sparklengths between 2 cm. diameter zinc spheres. It was determined by observations made with a coil the constants of which were known, so that the peak potential for any primary current was calculable. The smallest primary current which, at break, produced each sparklength was measured, and in the calculations allowance was made for the variation of the inductances with current. In any well constructed coil this variation is very small over a considerable range above and below the current at which the inductances have their maximum values, so that within this range (the primary and secondary capacities being supposed constant) the peak secondary potential is proportional to the primary current at break. The frequency of the slower oscillation of the coil used in the experiments was about 200 per sec., and the curve in Fig. 5 is suitable for the determination of the peak potential of any coil the principal oscillation of which has a frequency of about this value. Owing to the existence of time lag in the occurrence of sparks the sparking potentials indicated in Fig. 5 are rather greater than the steady potentials which would produce the same sparklengths. On the other hand, for potential waves of much greater frequency, the potential values given in Fig. 5 are too small. In working with small coils the writer has found very useful a gap in which one of the spheres has a very small hole bored axially in it, in the hole being placed a minute speck of radium bromide. This radium gap appears to have no time lag.*

Earlier Theories of the Action of Induction Coils. Early writers on the action of induction coils were content to regard

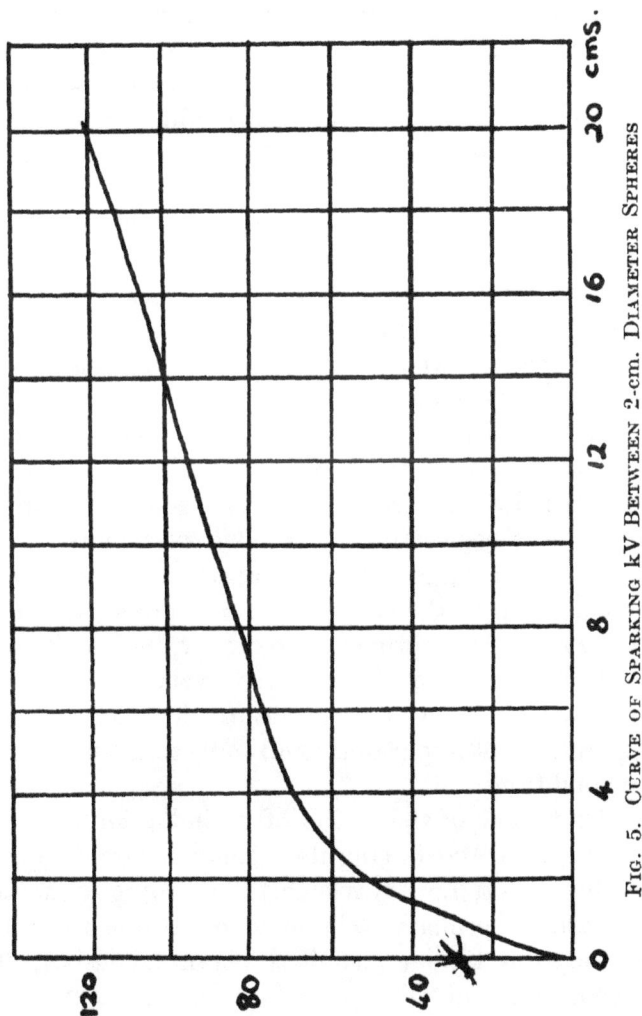

FIG. 5. CURVE OF SPARKING kV BETWEEN 2-cm. DIAMETER SPHERES

the high potential which appeared at the secondary terminals at "break" as arising from the diminution of the magnetic

* The lag may also be diminished by the use of a "third point." See J. D. Morgan, *Phil. Mag.* 4, p. 91, 1927 The action of the third point was traced by C. E. Wynn-Williams (*Phil. Mag.* 1, p. 353, 1926) to radiation from the point. The nature and source of the radiation have been carefully examined by J. Thomson, *Phil. Mag.* 5, p. 513; 6, p. 526, 1928; 7, p. 970; 8, p. 977, 1929.

flux in the iron core due to the primary current when this current was interrupted, and being proportional to the rate at which this flux could be caused to disappear. In this simple view of the matter the facts are ignored that the secondary winding possesses electrostatic capacity, that consequently there may be a considerable current in the secondary coil even though there is no discharge passing between its terminals, and that this secondary current also produces magnetic flux, a part (but not the whole) of which passes through the core. As a matter of fact, the rate of disappearance of the core-flux is in itself no criterion as to the magnitude of the maximum potential produced at the secondary terminals; in some cases, as we shall see later, the potential can be increased by prolonging the period of decay of the flux.

In 1891 a theory was proposed by the Russian physicist Colley,[*] in which the secondary potential was regarded as arising from the superposition of two oscillations. The frequencies of these oscillations were, however, assumed to be the separate circuit frequencies. The reaction of the secondary current on the primary was neglected, and in his experiments Colley expressly connected large self-inductances in series with the primary coil, so that the secondary acted inductively on only a small portion of the primary circuit, or, as we should now express it, so that the circuits were very loosely coupled. Colley's theory therefore only applies to the case of very loose coupling, and is not applicable to an induction coil in ordinary working conditions.

In 1901 a theory of the action of an induction coil, worked by an interrupter and having its secondary circuit open, was proposed by the late Lord Rayleigh.[†] According to this theory, the current in the primary coil is suddenly stopped at "break," while at the same time a current is instantaneously generated in the secondary coil. The secondary current then begins to oscillate as a system with one degree of freedom, the potential reaching its highest value at a quarter-period after the beginning of the oscillations. This is, at any rate, according to

---

[*] *Wied. Ann.*, 44, p. 109 (1891).

[†] *Phil. Mag.*, ii, p. 581 (1901). Rayleigh states that he regarded it as improbable that the current in the primary coil could also execute oscillations.

Rayleigh's theory, the ideal to be aimed at, but ordinary interrupters are supposed to be incapable of producing a sufficiently rapid break to satisfy the theoretical conditions unless the primary current is very small. It was pointed out by Rayleigh that there is a loss of energy at the moment of break except in the case when there is no magnetic leakage between the primary and secondary circuits, i.e. when $L_1 L_2 = M^2$, where L_1 and L_2 are the self-inductances of the circuits and M is their mutual inductance. Rayleigh showed that if this condition is satisfied and if the resistance of the secondary circuit is negligible, the principle of energy leads to the expression for the maximum secondary potential

$$V_{2m} = i_0 \sqrt{(L_1/C_2)}.$$

in which C_2 is the secondary capacity and i_0 is the primary current immediately before interruption. According to Rayleigh's theory, the maximum secondary potential is independent of the resistance of the primary circuit.

Another feature of the theory is that the only use of the primary condenser is to check the formation of an arc at the interrupter, and thus to quicken the break, so that when the interrupter is sufficiently rapid in action without this assistance the presence of the condenser is a disadvantage, since it then retards the decay of the primary current. In support of this conclusion, Rayleigh describes his experiments on the secondary spark-length of an Apps coil used with various forms of very rapid interrupter. In his final experiments he produced the break by severing the primary wire with a rifle bullet, and found that the secondary spark was longer without any condenser (or with a certain condenser of very small capacity) than when the usual coil condenser was connected across the broken part of the circuit. In other experiments, in which a less rapid form of interrupter, worked by a falling weight, was employed, he found that when the primary current was very weak longer secondary sparks were obtained without than with the coil condenser. A similar result has been recorded by W. H. Wilson,* who observed that when the magnetic energy supplied to an induction coil was insufficient to cause

* *Proc. Roy. Soc.*, LXXXVII, p. 76 (1912).

any appreciable spark at the interrupter, a longer secondary spark could be obtained without than with a primary condenser. No indication is given, however, as to whether the experiment was tried with various condensers of different capacities.

These experiments appear at first sight to afford strong evidence in favour of Rayleigh's theory, and the well-known fact that when an ordinary interrupter is used there is a certain most effective, or optimum, capacity—a fact which appears to contradict the theory—might be attributed to insufficient rapidity in the action of the interrupter.

On the other hand, there is now abundant experimental evidence of the existence of two oscillations, both in the primary and in the secondary circuit, and the oscillations in the primary are not taken into account in Rayleigh's theory. Further, Rayleigh's rifle-bullet experiment was repeated by the present writer with a very different result, viz. it was found that a decidedly longer spark could be obtained with than without a condenser provided the capacity of the condenser was suitably chosen.

It appears, therefore, that Rayleigh's theory, though applicable in a certain limiting case, is not sufficiently wide to cover the action of induction coils in general, in which the coupling falls well short of unity, and in which the oscillations of the primary current have an important influence on the phenomena exhibited.

The Present Theory. The theory, a short account of which is given in the earlier portion of the present chapter, and which will be developed in the next chapter, was suggested by the present writer in 1909. A full account of the theory was given in the author's *Theory of the Induction Coil** with a description of many experiments made with the object of testing the theory. In all cases the tests completely verified the theory.

It has already been indicated that the explanation of the action of induction coils and other oscillation transformers involves a detailed study of transient currents in coupled circuits. In these instruments, in fact, the whole action depends upon transient currents, which are utilized in them for the production of certain effects. It is different in most of the

* London, Sir Isaac Pitman & Sons (1921).

applications of electricity, in which use is made of constant or of steadily alternating currents, and the importance of surges is often under-estimated, sometimes with destructive results, as when current is switched off too suddenly from a highly inductive circuit. Sometimes also the neglect of transitional effects is apt to lead to erroneous conclusions in regard to fundamental matters, the following being a case in point.

The Law of Electromagnetic Induction. It is well known that the law which describes the conditions necessary for the production of induced currents is stated in two ways, viz. (1) a current is induced in a wire (forming part of a closed circuit) when lines of magnetic induction are cutting across it, (2) a current is induced in a circuit when the total number of lines of induction passing through the circuit is changing. In many cases the two statements are equivalent, the change in the number of lines linked with the circuit being equal to the number which cut across the wire in the same time. But there are cases in which lines cut across a conductor forming part of a circuit without causing change of the total number linked with the circuit, and other cases in which the total number of lines linked with a circuit changes without any of them cutting across the conductor. The following argument is sometimes advanced to show that the cutting of lines by the wire is not a necessary condition. Suppose that a circular ring of uniform section is uniformly wound with wire, forming a circular solenoid, the adjacent ends of the wire being connected to a battery and a contact breaker. Another wire, forming a closed secondary circuit, is linked with the ring. Then, if current flows in the solenoid its magnetic field is entirely confined to the space within the ring and, therefore, if the current is interrupted, none of the lines of induction will pass outside the primary coil so as to be able to cut across the secondary wire. Nevertheless, the interruption of the primary current gives rise to an induced current in the secondary circuit.

Now it is true that if the current in the solenoid is constant in time its value is uniform round the ring, and it produces no magnetic field outside the ring, but the fallacy in the above argument is seen when we consider the transitory effect of the interruption of the primary current. The solenoid possesses both capacity and self-inductance. Immediately after break an oscillation is set up, the current during the oscillation being zero (or nearly so) at the part of the solenoid where the leads are connected and a maximum at the diametrically opposite part. The oscillating current is not uniformly distributed round the ring, its field is not confined to the space within it, and its lines of induction spread out into the surrounding space and cut across the secondary wire. The same is true if the primary current is varied in any other manner.

This example, therefore, forms no exception to the first statement of the law, substantially the statement given by Faraday in 1831, which will probably be found to be true in all cases if transient effects are fully taken into account.

CHAPTER II

THE MAXIMUM SECONDARY POTENTIAL

PROCEEDING now with the discussion of the expressions (21) and (22), page 12, which give the values of the secondary and primary potential of an induction coil at any time after break, we will first consider the maximum value of the secondary potential and the manner in which this maximum can be made as great as possible, still confining our attention to the case in which the resistances of the circuits and other causes of damping (e.g. secondary discharge) are neglected.

The expression (21) for V_2 represents two superposed oscillations which begin in opposite phase, and the amplitudes of which are inversely as the frequencies. In actual coils, owing to the damping of the oscillations, the peak potential in the second and following half waves of the slower oscillation is always considerably less than that in the first half wave, so that in considering the maximum potential we may confine our attention to the first half wave of the slower component.

Sum of Amplitudes. The first point to be noticed is that the value of V_2 cannot be greater than the sum of the amplitudes of the two components. This sum is easily seen by (21) to be equal to

$$2\pi L_{21}\, i_0 \, \frac{n_1 n_2}{n_2 - n_1},$$

which represents the greatest conceivable secondary potential of an induction coil, only attainable on the supposition that all causes of damping are neglected. It may also be expressed in terms of the constants of the circuits. Employing the expression (13) for the frequencies, and setting $u = L_1 C_1 / L_2 C_2$ as before, we find that the sum of the amplitudes is equal to

$$\frac{L_{21}\, i_0}{\sqrt{(L_2 C_2)}} \cdot \frac{1}{\sqrt{[1 + u - 2\sqrt{\{u(1 - k^2)\}}]}} \qquad . \qquad . \qquad (26)$$

or $\dfrac{L_{21}\, i_0}{\sqrt{(L_2 C_2)}}\, U$, if $U^2 = \dfrac{1}{1 + u - 2\sqrt{\{u(1 - k^2)\}}}$ $\qquad . \qquad . \qquad (27)$

In terms of the reciprocal ratio $m(= L_2C_2/L_1C_1)$ the expression for the sum of the amplitudes is—

$$\frac{L_{21}\,i_0}{\sqrt{(L_1C_1)}}\,M \qquad . \qquad . \qquad . \qquad . \qquad . \qquad . \qquad . \qquad (28)$$

where $M^2 = \dfrac{1}{1 + m - 2\sqrt{\{m(1-k^2)\}}}$ . . . . (29)

Best Frequency Ratios. The next point is that the two component potential waves in the secondary can only conspire to

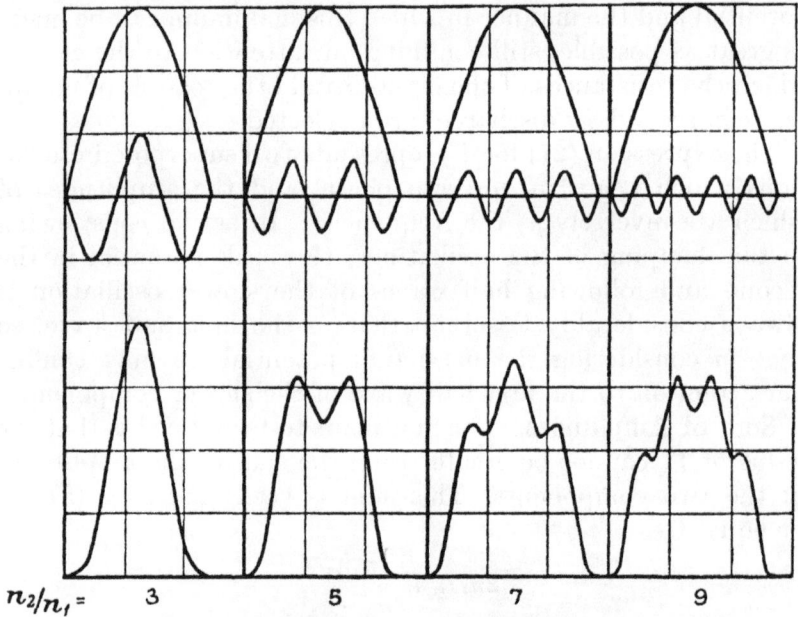

$n_2/n_1 =$ 3   5   7   9

FIG. 6. OSCILLATIONS IN SECONDARY OF INDUCTION COIL

produce a maximum equal to the sum of their amplitudes when the frequency ratio has certain values. In Fig. 6, the upper curves show the two component waves in the secondary for four frequency ratios. Only one half wave of the slower component is represented, the components have amplitudes inversely as their frequencies, and they begin in opposite phase. The lower curves show the result of the superposition of the components, and, therefore, the form of the resultant potential wave in the secondary circuit.

It will be seen from Fig. 6 that positive maxima of the two

components conspire when the frequency ratio is 3 or 7, and a little further consideration shows that the same happens when the frequency ratio has one of the values 11, 15, 19, . . . For any one of these values of the frequency ratio, therefore, the maximum secondary potential is equal to the sum of the amplitudes of the component oscillations, and is, therefore, represented by either of the expressions (26) or (28). If the frequency ratio has any other value the maximum potential falls short of the sum of the amplitudes, the deficiency being greatest at the ratios 5, 9, 13, . . . at any one of which a minimum of the rapid oscillation occurs simultaneously with a maximum in the slower component, as shown in the second and fourth curves of Fig. 6.

Most Effective Adjustments. There is still another condition to be satisfied in order to ensure that the secondary potential shall attain its greatest possible value, for the condition $n_2/n_1 = 3, 7, 11, . . .$, though necessary, is not sufficient for this purpose. The expression (26) shows that the sum of the amplitudes depends upon the value of the ratio u, and that it can be varied by changing the primary capacity (which is proportional to u) without altering any of the other constants of the circuits. By differentiating the expression (27) for U^2 with respect to u we find that this quantity has a maximum value of $1/k^2$ when $u = 1 - k^2$. Consequently, if the two conditions—

$$u = 1 - k^2 \qquad . \qquad . \qquad . \qquad . \qquad . \qquad . \qquad . \qquad . \qquad (30)$$

$$\frac{n_2}{n_1} = 3, 7, 11, 15, . . . \qquad . \qquad . \qquad . \qquad . \qquad . \qquad (31)$$

are satisfied, the maximum secondary potential V_{2m} has its greatest possible value, which is given by—

$$V_{2m} = \frac{L_{21} i_0}{\sqrt{(L_2 C_2)}} \cdot \frac{1}{k} \qquad . \qquad . \qquad . \qquad . \qquad . \qquad (32)$$

$$= i_0 \sqrt{\frac{L_{21}}{L_{12}}} \sqrt{\frac{L_1}{C_2}}, \qquad . \qquad . \qquad . \qquad . \qquad (33)$$

so that

$$\tfrac{1}{2} L_1 i_0^2 = \tfrac{1}{2} C_2 V_{2m}^2 \cdot \frac{L_{12}}{L_{21}} \qquad . \qquad . \qquad . \qquad . \qquad . \qquad (34)$$

The last equation (34) is the energy equation for an induction coil in one of the adjustments specified by (30) and (31). It expresses that in these adjustments the whole of the magnetic energy $\frac{1}{2}L_1 i_0^2$ initially supplied to the primary circuit becomes converted soon after break into electrostatic energy in the secondary, and therefore gives rise to the greatest possible value of the secondary potential for a given primary current i_0 at break. It is only in these favourable adjustments that this complete conversion of the energy takes place.

By numerical calculation from equations (30), (31), and (14), the last of which gives n_2/n_1 in terms of u and k^2, these adjustments can be completely specified. The first four of the series are given in Table 1, in which the first column contains the values of the frequency ratio, the second those of the coupling, and the third the values of the ratio u, that is $L_1 C_1/L_2 C_2$.

TABLE 1

n_2/n_1	k^2	$\dfrac{L_1 C_1}{L_2 C_2} = 1 - k^2$
3	0·571	0·429
7	0·835	0·165
11	0·902	0·098
15	0·931	0·069

If the coupling has one of the values given in the second column of Table I, the optimum value of u is given by the corresponding number in the third column, and the most effective value of the primary capacity is then $(1 - k^2)L_2 C_2/L_1$.

The adjustments specified in Table I, giving complete conversion of magnetic into electrostatic energy, are the only four which are likely to occur in the working conditions of actual coils. In the first, the coupling 0·571 is smaller than that usually found in coils, but this value can be easily obtained by inserting series inductance in the primary circuit, or, in some coils, by drawing out the primary to a suitable distance along the axis of the secondary. The fourth value 0·931 may perhaps be found in coils in which the core forms a nearly closed iron circuit, especially if the secondary terminals are connected with a condenser. In actual coils the energy conversion is, of course, incomplete owing to losses. Nevertheless, experimental evidence, so far as it has been obtained, shows that the conversion is a maximum in the adjustments of Table I.

The Primary Potential. The physical meaning of the condition $u = 1 - k^2$ may be further illustrated by considering what goes on in the primary circuit after "break." In this circuit the two oscillations begin in the *same* phase, so that if the frequency ratio has one of the values 3, 7, 11, . . . the primary potential at the instant $t = 1/4n_1$ (at which moment the secondary potential reaches its greatest value) is equal to the *difference* of the amplitudes. The amplitudes of the potential oscillations in the primary are, by (22)—

$$\frac{2\pi i_0\, n_1 n_2^2}{C_1(n_2^2 - n_1^2)}\left(L_1 C_1 - \frac{1}{4\pi^2 n_2^2}\right),$$

and

$$-\frac{2\pi i_0\, n_1^2 n_2}{C_1(n_2^2 - n_1^2)}\left(L_1 C_1 - \frac{1}{4\pi^2 n_1^2}\right),$$

and their difference is

$$2\pi i_0\, L_1 \frac{n_1 n_2}{n_2 - n_1} - \frac{i_0}{2\pi C_1}\cdot\frac{1}{n_2 - n_1}$$

$$= \frac{i_0}{2\pi C_1(n_2 - n_1)}\left(4\pi^2 L_1 C_1\, n_1 n_2 - 1\right)$$

which, by (13), reduces to

$$\frac{i_0}{2\pi C_1(n_2 - n_1)}\left(\sqrt{\frac{u}{1 - k^2}} - 1\right).$$

The condition $u = 1 - k^2$ makes this difference vanish, so that, in the adjustments of Table I, the primary condenser is without charge at the moment of maximum secondary potential.

In Fig. 7, the upper curves represent the two components of the primary potential oscillations for three frequency ratios, the amplitudes being equal as required by the condition $u = 1 - k^2$. The lower curves show the result of their superposition. It will be seen from the lower curves that if the frequency ratio is 3 or 7 the primary potential is zero, and, therefore, the condenser is uncharged, at the time $t = 1/4n_1$

(a quarter period of the slower oscillation) after break. The curves show, moreover, that not only V_1 but also its rate of variation dV_1/dt is zero at this moment, so that, since $i_1 = C_1 \, dV_1/dt$, there is also no current in the primary circuit.

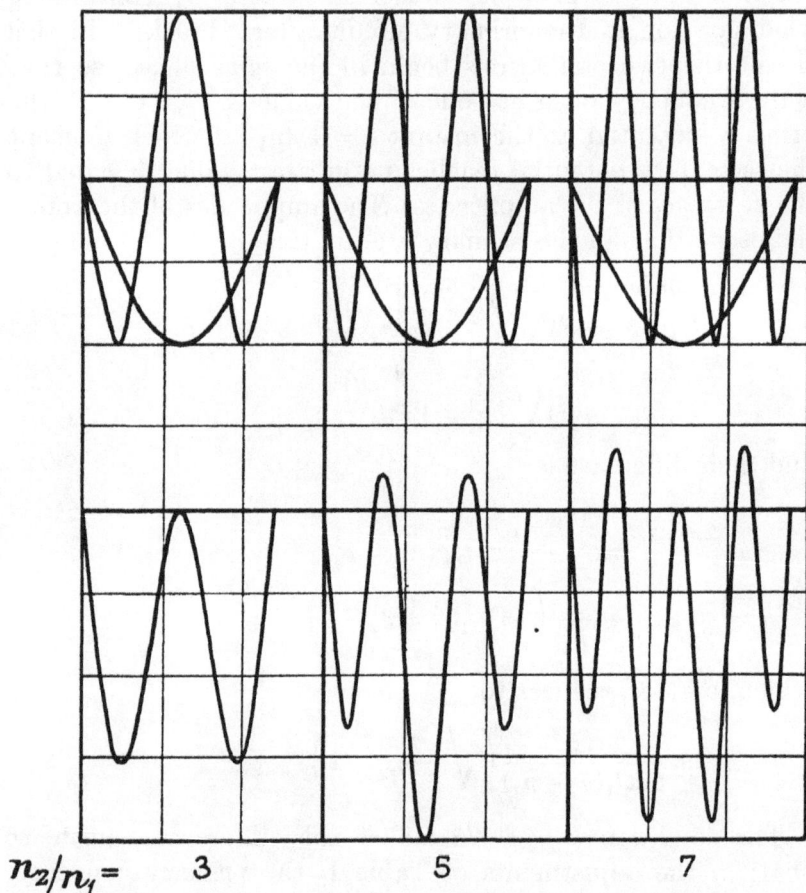

$n_2/n_1 =$ 3   5   7

FIG. 7. OSCILLATIONS IN PRIMARY OF INDUCTION COIL

In the adjustments of Table I there is, therefore, neither magnetic nor electrostatic energy in the primary circuit at the moment of maximum secondary potential. At this instant also, since $dV_2/dt = 0$, there is no current in the secondary circuit. The whole of the energy therefore exists in the electrostatic form in the secondary, as already indicated by the energy equation (34).

The Maximum Secondary Potential in any Adjustment. If the coupling has not one of the special values given in Table I, positive maxima in the two waves of secondary potential do not occur simultaneously, and the maximum is necessarily less than the sum of the amplitudes. To find its value we proceed as follows: From the expression (21) for V_2 we find the turning points in the (V_2, t) curve, determined by $dV_2/dt = 0$, to occur at times given by

$$\cos 2\pi n_1 t - \cos 2\pi n_2 t = 0, \qquad . \qquad . \qquad . \qquad . \qquad (35)$$

or $\quad \sin \pi(n_1 + n_2)t \cdot \sin \pi(n_2 - n_1)t = 0.$

The stationary values of V_2 therefore occur at the times

$$\left. \begin{aligned} t &= 0, \; \frac{1}{n_1 + n_2}, \; \frac{2}{n_1 + n_2}, \; \frac{3}{n_1 + n_2}, \qquad . \qquad . \\ \text{and} \quad & \frac{1}{n_2 - n_1}, \; \frac{2}{n_2 - n_1}, \; \frac{3}{n_2 - n_1}, \qquad . \qquad . \end{aligned} \right\} \qquad . \quad (36)$$

At any stationary value we have, by (35),

$$\sin 2\pi n_2 t = \pm \sin 2\pi n_1 t, \qquad . \qquad . \qquad . \qquad . \qquad (37)$$

the upper sign giving the numerical minima of V_2, the lower sign the maxima.

Substituting in (21) we find that all the maximum values of V_2 are given by the simple expression—

$$2\pi L_{21} i_0 \frac{n_1 n_2}{n_2 - n_1} \sin 2\pi n_1 t, \qquad . \qquad . \qquad . \qquad . \qquad (38)$$

t having for each maximum the appropriate value given in the upper line of (36). Thus, the first maximum in the (V_2, t) curve occurs at time $t = \dfrac{1}{n_1 + n_2}$, the second at $t = \dfrac{2}{n_1 + n_2}$, and so on.

The form of the expression (38) shows that the maxima all lie on the sine curve

$$V = 2\pi L_{21} i_0 \frac{n_1 n_2}{n_2 - n_1} \sin 2\pi n_1 t \qquad . \qquad . \qquad . \qquad (39)$$

as illustrated by the broken line curve in Fig. 8 which passes through the maxima of the (V_2, t) curve represented by the full line.

The greatest value of the secondary potential is the maximum which occurs nearest to the summit of the sine curve (39). It may be called the *principal maximum* in the (V_2, t) curve. It is easily seen that the first maximum, occurring at time $\dfrac{1}{n_1 + n_2}$, is the principal maximum if the frequency

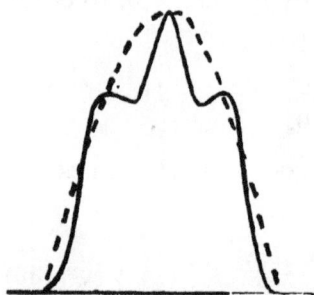

FIG. 8. SHOWING MAXIMA LYING ON A SINE CURVE

ratio n_2/n_1 is between 1 and 5. If $n_2 = 5n_1$ the first maximum is equal to the second, and they occur at times $\dfrac{1}{n_1 + n_2}, \dfrac{2}{n_1 + n_2}$. If n_2/n_1 is between 5 and 9, the second maximum is the principal maximum, occurring at the time $\dfrac{2}{n_1 + n_2}$. If $n_2 = 9n_1$ the second and third maxima are equal, and if n_2/n_1 is between 9 and 13 the third maximum is the principal maximum, and it occurs at

$$t = \frac{3}{n_1 + n_2} \text{ ; and so on.}$$

Consequently, the principal maximum secondary potential is given by the equation

$$V_{2m} = 2\pi L_{21} i_0 \frac{n_1 n_2}{n_2 - n_1} . \sin \phi, \qquad . \qquad . \qquad . \qquad (40)$$

where

$$\left.\begin{array}{l} \phi = \dfrac{2\pi n_1}{n_1 + n_2} \text{ if } \dfrac{n_2}{n_1} \text{ is between 1 and 5,} \qquad . \\[3mm] \text{,, } \quad \phi = \dfrac{4\pi n_1}{n_1 + n_2} \quad \text{,,} \qquad \text{,,} \qquad 5 \text{ ,, } 9, \qquad . \\[3mm] \text{,, } \quad \phi = \dfrac{6\pi n_1}{n_1 + n_2} \quad \text{,,} \qquad \text{,,} \qquad 9 \text{ ,, } 13, \qquad . \end{array}\right\} \quad . \quad (41)$$

and so on.

If n_2/n_1 has one of the values 3, 7, 11, . . . the maxima of the two oscillations occur simultaneously, the principal maximum occurring at the time $1/4n_1 \left(\phi = \dfrac{\pi}{2}\right)$, that is, it occurs at the summit of the sine curve (39). Its value in such cases is, by (40), $2\pi L_{21} i_0 \dfrac{n_1 n_2}{n_2 - n_1}$, which is equal to the sum of the amplitudes (26).

In any adjustment, however, the principal maximum is given by (40) which, by (27), is equivalent to

$$V_{2m} = \frac{L_{21} i_0}{\sqrt{(L_2 C_2)}} U \sin \varphi, \qquad . \qquad . \qquad . \qquad . \qquad (42)$$

the angle φ being given by (41).

Efficiency of Conversion. Equation (42) may be written in the form

$$\tfrac{1}{2} C_2 V_{2m}{}^2 \frac{L_{12}}{L_{21}} = \tfrac{1}{2} L_1 i_0{}^2 k^2 U^2 \sin^2\varphi \qquad . \qquad . \qquad . \qquad . \qquad (43)$$

giving the maximum value of the electrostatic energy in the secondary circuit. The ratio of this quantity to the initial primary magnetic energy $\tfrac{1}{2} L_1 i_0{}^2$ is $k^2 U^2 \sin^2\varphi$. This ratio may be called the "efficiency of conversion" of the coil. In the special adjustments of Table I its value is, of course, unity.

Optimum Primary Capacity. The optimum primary capacity may be found by calculation if the primary self-inductance L_1, the coupling k^2, and the product $L_2 C_2$ of self-inductance and capacity for the secondary coil are known. We first find the value of u which, for given k^2, gives the greatest value of $U \sin \varphi$. To do this we insert the value of k^2, with any value of u (not far from $1 - k^2$) in equation (14), and hence determine the frequency ratio n_2/n_1. Then we calculate φ from (41) and U from (27), and thus obtain the value of $U \sin \varphi$. Repeating the process with various values of u we eventually find the value which gives the maximum of $U \sin \varphi$, i.e. the optimum value of u for the given coupling. In Table II are collected the optimum values of u (second column) for various values of k^2 (first column) covering practically the whole of the useful range. The third column shows the values of the frequency ratio at which the greatest maxima of $U \sin \varphi$ occur, the fourth and fifth columns the corresponding values of U and $U \sin \varphi$, and the last column the values of the efficiency of conversion.

TABLE II

ADJUSTMENTS FOR MAXIMUM SECONDARY POTENTIAL

k^2.	$u\left(=\dfrac{L_1 C_1}{L_2 C_2}\right)$	$\dfrac{n_2}{n_1}$.	U.	$U \sin \varphi$.	Efficiency of Conversion. $k^2 l^{\cdot 2} \sin^2 \varphi$.
0·92	0·0628	15	1·042	1·042	0·999
0·92	0·1297	11	1·040	1·040	0·995
0·90	0·0965	11	1·054	1·054	1·0
0·87	0·075	10·79	1·067	1·067	0·990
0·87	0·21	7·183	1·066	1·066	0·989
0·835	0·165	7·0	1·094	1·094	1·0
0·768	0·11	6·801	1·125	1·124	0·970
0·71	0·09	6·595	1·142	1·138	0·919
0·71	0·44	3·767	1·174	1·137	0·918
0·70	0·445	3·68	1·183	1·153	0·930
0·64	0·45	3·299	1·246	1·238	0·981
0·60	0·435	3·12	1·290	1·288	0·996
0·571	0·429	3·0	1·323	1·323	1·0
0·5	0·41	2·752	1·408	1·401	0·981

For any value of k^2 given in Table II the optimum primary capacity is found by multiplying the corresponding value of u (second column) by $L_2 C_2/L_1$, and the principal maximum of V_2 is found by multiplying the value of $U \sin \varphi$ (fifth column) by $L_{21} i_0/\sqrt{(L_2 C_2)}$.

The numbers in the second column of Table II are the optimum values of the ratio $L_1 C_1/L_2 C_2$ for the degrees of coupling given in the first column. For intermediate values of k^2 within the same range the optimum values of u may be found approximately by plotting the (k^2, u) curve from the values given in the Table. This curve has three distinct segments, one covering very approximately the range $k^2 = 0\cdot92$ to $0\cdot87$, another the range $0\cdot87$ to $0\cdot71$, and the third running from $0\cdot71$ downwards. At each of these three values of k^2 there are practically two equal greatest maxima of $U \sin \varphi$, one corresponding to a value of u greater than, the other less than $1 - k^2$. For example, when k^2 is $0\cdot71$ the two values $0\cdot09$ and $0\cdot44$ of u give practically equal values of $U \sin \varphi$, these being higher than those corresponding to any other value of u. At this coupling therefore there are two optimum primary capacities.

At those values of k^2 which give two equal greatest maxima of $U \sin \varphi$, the greatest possible conversion efficiency (see Table II, last column) is less than it is at neighbouring values of the coupling. Of the "optimum" adjustments within the range of Table II the efficiency of conversion is least at $k^2 = 0.71$. In this case rather over 8 per cent of the initial energy supply $\frac{1}{2}L_1 i_0^2$ appears as electrostatic energy in the primary condenser and electrokinetic energy in the primary circuit at the moment when the secondary potential reaches its greatest value. It can be shown by (22) that of this 8 per cent about one-quarter represents the energy of charge of the condenser, the remainder the energy of the primary current.

The efficiency of conversion is, as explained above, unity in those adjustments in which the greatest maximum of $U \sin \varphi$ occurs at a frequency ratio having one of the values 3, 7, 11, . . . In Table II there are three such adjustments, giving the ratios 3, 7, and 11; they are the first three adjustments of Table I.

A very simple approximate rule may be stated for the optimum primary capacity if k^2 is between 0.71 and 0.5. It will be seen from Table II that there is within these limits no great variation in the optimum value of u. It should also be noted that, for such values of k^2, $U \sin \varphi$ varies very slowly with u near the greatest maximum; these portions of the $(u, U \sin \varphi)$ curves are very flat-topped. We may therefore, without much error, take the mean value of the optimum u as applicable over this range, and say that if k^2 lies between 0.71 and 0.5 the optimum primary capacity is that which makes $L_1 C_1 = 0.43 L_2 C_2$.

There is no such simple rule for the higher degrees of coupling, but it will be noticed in Table II that at these the greatest maxima of $U \sin \varphi$ correspond very closely to adjustments in which φ is equal to $\pi/2$. This applies with considerable accuracy down to the value $k^2 = 0.768$, and with fair accuracy down to 0.71. At the latter degree of coupling the value of $U \sin \varphi$ for the ratio 7 (approximately 1.134 at $u = 0.078$) only differs by 4 parts in 1,100 from the neighbouring maximum value (1.138) given in Table II. We may therefore state, as an approximate result, that if k^2 is greater than 0.71 the optimum primary capacity is that which makes the frequency ratio have that one of the values 3, 7, 11, . . . which most nearly makes U a maximum. At $k^2 = 0.71$ the approximate rule, therefore, gives the maximum value of

$U \sin \varphi$ with considerable accuracy, but not so accurately the optimum value of u, viz. $0 \cdot 078$ as against $0 \cdot 09$ (see Table II).

Although the optimum values of u given in Table II are here deduced from the simpler theory in which the oscillations are regarded as undamped, they are found to show good agreement with experimental results so far as these have been obtained. Some of these comparisons with experiment will be referred to later in the present chapter.

Capacity—Potential Curves.* We will now consider the manner in which the principal maximum secondary potential V_{2m} of an induction coil changes when the capacity of the primary condenser C_1 is varied, the coupling and the primary current at break being constant. In these circumstances V_{2m} is proportional to $U \sin \varphi$, the other factors in the expression (42) being constant, and C_1 is proportional to u.

As u is increased continually from zero to unity the frequency ratio, by (14), diminishes from infinity to a minimum of $\sqrt{(1 + k)/(1 - k)}$. The factor U, by (27), increases from unity and passes through a single maximum value (viz. $1/k$) corresponding to the relation $u = 1 - k^2$. The factor $\sin \varphi$, however, reaches its maximum value unity whenever the frequency ratio passes through one of the values . . . 15, 11, 7, 3.

Consequently, it might be expected that the capacity—potential curves. or the $(u, U \sin \varphi)$ curves, would show a series of maxima and minima as the primary capacity is increased from zero. The curves consist, in fact, of a series of arches the relative proportions of which depend upon the coupling. The curves may be calculated by determining the value of $U \sin \varphi$ for any values of u and k^2 in the manner already explained.

Five of these curves are shown in the following diagrams, in each of which the full-line curve represents $U \sin \varphi$, the broken-line curve U. The ordinate of the U curve is proportional to the sum of the amplitudes of the components of the potential wave in the secondary coil; it therefore represents what the maximum potential would be if the two components were always in the same phase at the moment of maximum

* *The Electrician*, 30th Aug., p. 376; 6th Sept., p. 396; 13th Sept., p. 416 (1918). See also *Phil. Mag.*, pp. 229. 230, Aug., 1915; p. 156, Aug., 1918

potential. The difference of the ordinates of the U and $U \sin \varphi$ curves represents the deficiency in the maximum potential arising from the fact that the two component oscillations are not generally in phase with each other at the instant when the maximum occurs.

The curves are drawn for values of u up to unity; their continuations beyond this range are of no importance in the theory of the optimum primary capacity, but they are of interest in connection with that of the optimum *secondary*

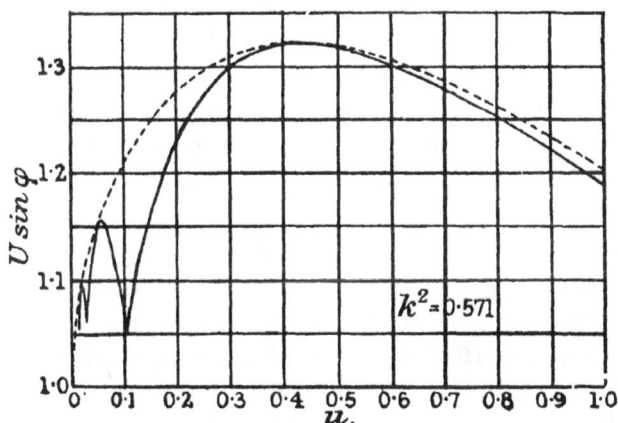

FIG. 9. VARIATION OF SECONDARY POTENTIAL WITH PRIMARY CAPACITY (CALCULATED) AT COUPLING 0·571

capacity, which will be discussed later. In each case the U curve passes through the point $(0, 1)$, since the value of U is unity when u is zero.

The first diagram (Fig. 9) refers to the case in which k^2 has the value 0·571, the smallest of the values of the coupling giving unit efficiency of conversion (see Table I, p. 32). It will be seen that the $U \sin \varphi$ curve consists of a series of arches, all lying within and each touching the U curve. These may be called the 3/1, 7/1, 11/1, ... arches, since their points of contact with the U curve occur at these values of the frequency ratio. The maximum value of U occurs at $u = 0.429$ $(= 1 - k^2)$, and the 3/1 arch touches the U curve at its highest point. The maximum of U and the greatest maximum of $U \sin \varphi$, each of which has the value 1·323, occur at the same value of u, and this is the condition for maximum efficiency of conversion of primary magnetic into secondary electrostatic energy.

The points of intersection of the arches, i.e. the minimum points in the $U \sin \varphi$ curve, correspond with the values 5, 9, 13, . . . of the frequency-ratio, the three shown in the diagram occurring at $u = 0 \cdot 105$, $0 \cdot 03$, and $0 \cdot 014$. The corresponding values of the angle φ are $\pi/3$ (or $2\pi/3$), $2\pi/5$ (or $3\pi/5$), $3\pi/7$ (or $4\pi/7$). These points of intersection represent adjustments in which there are two equal principal maxima in the (V_2, t) curve; at the first, the intersection of the $3/1$ and $5/1$ arches, the first and second maxima are equal; at the second, the second and third maxima, and so on. The arches may be produced below the points of intersection, but these lower portions of the curves represent *secondary* maxima in the (V_2, t) curve.

If produced to the left, i.e. towards the origin, the $U \sin \varphi$ curve would disclose an infinite series of diminishing arches, touching the U curve at points corresponding with the frequency-ratios 15, 19, 23, . . . and meeting each other at the ratios 17, 21, 25, . . . Since these arches all lie within (or below) the U curve, it is clear that when $u = 0$ (i.e. $C_1 = 0$) the value of $U \sin \varphi$ cannot be greater than unity. According to the present theory, therefore, the greatest maximum value of the secondary potential when $k^2 = 0 \cdot 571$, which occurs at $u = 0 \cdot 429$, is $1 \cdot 323$ times the value attainable when $C_1 = 0$, even on the supposition that the interruption is perfectly sudden.

The curves for other degrees of coupling show the same general characteristics, but there are important differences in detail. Fig. 10 represents the case $k^2 = 0 \cdot 71$, which gives very approximately two equal greatest maxima of $U \sin \varphi$. In this case the $3/1$ arch has no real point of contact with the U curve, since the greatest value of k^2 which allows the existence of the $3/1$ ratio is $0 \cdot 64$. The summits of the $3/1$ and $7/1$ arches have practically equal ordinates, each being, however, well below the maximum of U. There are thus two "optimum" capacities, given by $u = 0 \cdot 44$ and $0 \cdot 09$, but even with these capacities the conversion efficiency is only $0 \cdot 918$. It is clear, therefore, that the coupling $k^2 = 0.71$ is disadvantageous from the point of view of efficiency of conversion.

The 5/1 minimum occurs at $u = 0.176$, the value of $U \sin \varphi$ being 1·017. In this adjustment the conversion efficiency is

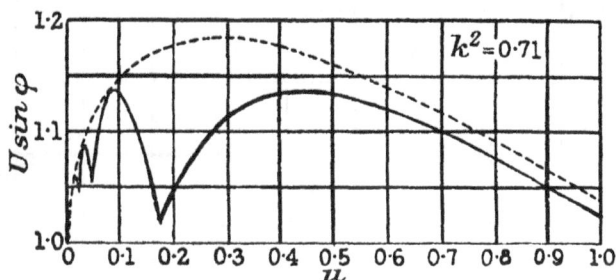

FIG. 10. VARIATION OF SECONDARY POTENTIAL WITH PRIMARY CAPACITY (CALCULATED) AT COUPLING 0·71

only 0·734. It can be shown, by calculation from the expression (25) for the core flux, that the time in which the flux falls to zero in the adjustment $k^2 = 0.71$, $u = 0.176$, is only 0·82 of the time required in the adjustment corresponding to the

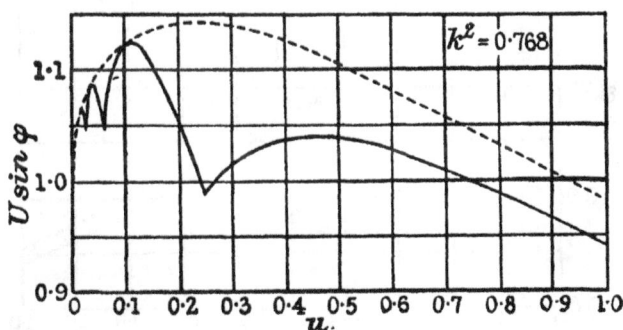

FIG. 11. VARIATION OF SECONDARY POTENTIAL WITH PRIMARY CAPACITY (CALCULATED) AT COUPLING 0·768

summit of the first arch of Fig. 10. Thus, the less rapid disappearance of flux is accompanied by a higher secondary potential, a result which is at variance with the earlier theories of induction coil action.

Fig. 11 shows the curves for $k^2 = 0.768$, the coupling found by the writer for a certain coil. In this case the greatest value of $U \sin \varphi$ (1·124) occurs in the 7/1 arch, near its contact with the U curve. The optimum capacity is given by $u = 0.11$, the frequency-ratio being 6·801. The 5/1 minimum occurs at

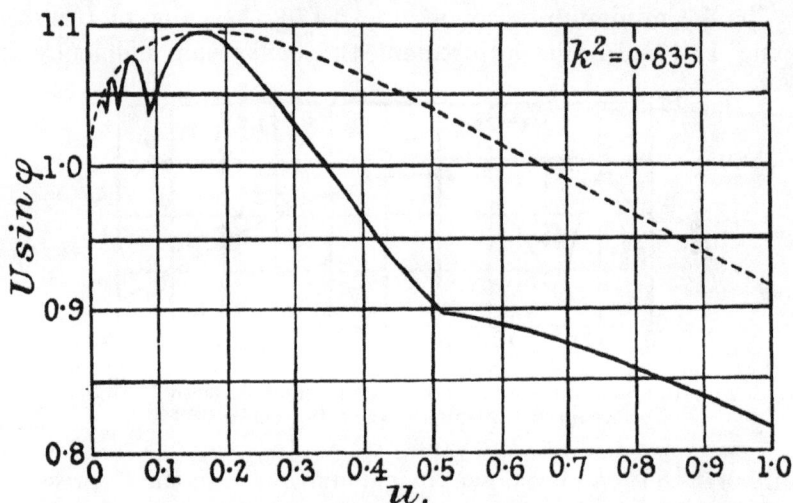

FIG. 12. VARIATION OF SECONDARY POTENTIAL WITH PRIMARY CAPACITY (CALCULATED) AT COUPLING 0·835

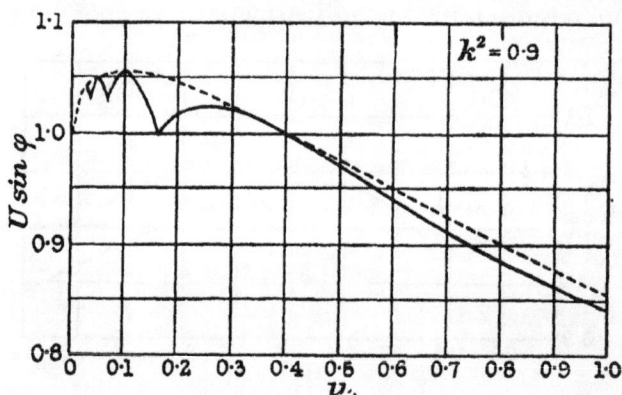

FIG. 13. VARIATION OF SECONDARY POTENTIAL WITH PRIMARY CAPACITY (CALCULATED) AT COUPLING 0·9

$u = 0.248$, and the 3/1 arch has a maximum of 1·038 at $u = 0.46$.

Figs. 12 and 13 represent the curves for $k^2 = 0.835$ and $k^2 = 0.9$, which are very approximately the second and third "unit-efficiency" values of the coupling. In each of these cases the U curve is touched at its highest point by the $U \sin \varphi$ curve; in Fig. 12 by the summit of the 7/1 arch, in Fig. 13

by that of the 11/1 arch. The optimum values of u for these cases are 0·165 and 0·0965,* these being practically equal to the corresponding values of $1 - k^2$. In Fig. 13 there is no trace of the 3/1 arch, the smallest value of the frequency-ratio (viz. $\sqrt{[(1 + k)/(1 - k)]}$ at $u = 1$) being 6·163, which exceeds the 5/1 limit of this arch.

From these examples it will be seen that while the theoretical capacity-potential curves for different values of k^2 have the same general features, they differ from one another in certain important characteristics, such as the relative heights of the various arches, the ratio of the greatest maximum of $U \sin \varphi$ to the neighbouring minimum on the left side of it, the value of u at which the 5/1 minimum or the 7/1 maximum occurs, and so on. To each value of k^2 corresponds a certain form of capacity-potential curve, so that if the curves could be obtained by experiment, and if the characteristics of the curves are not too greatly modified by the resistances, some idea might be obtained from them as to the coupling of the circuits. The curves given above refer, as already stated, to the case of an ideal induction coil in which the resistances are negligible and in which the interruption of the primary current takes place with perfect suddenness. To what extent the form of the curves is modified in the working conditions of an actual coil will be seen in the examples of experimental curves given in the next section.

Experimental Capacity—Potential Curves. The experimental curves may be determined for any coil without the use of any other measuring instruments than an amperemeter and an adjustable spark gap. A battery, a rheostat, and a variable condenser are required for the primary circuit. One method of procedure is to find the greatest secondary spark-length for a given primary current with various capacities connected across the interrupter. A better plan, however, is to keep the distance between the spark terminals constant and determine for each value of the capacity the least primary current which will cause the spark to pass. This method is much more convenient than the other, since it allows readings to be taken

---

* The exact value of the third "unit-efficiency" coupling is rather greater than 0·9, being more nearly equal to 0·9025.

more quickly; and, further, since the secondary potential of an induction coil is approximately proportional to the primary current just before interruption—so long as the current is not varied beyond certain limits--and since in this method the secondary potential is constant, the reciprocal of the minimum sparking current may (except in certain cases which will be referred to later) be taken as a measure of the maximum secondary potential developed at the interruption of a given primary current. It is thus unnecessary to know the relation between the spark-length and the sparking potential. This method of the constant secondary potential was therefore adopted in determining the curves shown below, in which the reciprocal of the least sparking current is plotted against the primary capacity.

In order that the experiments should include a considerable number of values of $u(= L_1C_1/L_2C_2)$ between 0 and 1, it is desirable to connect the secondary terminals with a condenser. This increases L_2C_2 and therefore allows the smaller values of u to be attained without the employment of very small primary capacities which might fail to prevent sparking at the interrupter.

The curves shown were obtained with an 18 in. coil, the primary of which was movable along the axis of the secondary so that the coupling could be varied in this way. Methods for determining the coupling are described in the next chapter. The secondary terminals were connected with a spark gap having spherical zinc electrodes and, for the first four curves, with a condenser of about 0·001 mfd. capacity. The interrupter was of the hand-operated mercury dipper variety. In the first four experiments the gap was set at 2·31 mm. (maximum V_2 about 9,000 volts), the primary current required to produce the spark being about 1 ampere. The primary current measured (i_0) was that which caused the sparks to appear at one half the number of interruptions. It was generally found that the current could be so adjusted that the sparks passed for considerable intervals at alternate interruptions, a slight variation of the current one way or the other causing the sparks to fail altogether or to appear at every break. Preliminary experiments showed where the maxima and minima on the curves lay, and after these the points were determined in the order in which they occur on the curves, beginning with the largest capacity, and without repetitions. The points marked ○ were the actual points observed. Above each diagram are indicated some experimental values of u ($= L_1C_1/L_2C_2$) the values of L_1 and L_2C_2 being determined by methods to be described in the next chapter.

In the first experiment (Fig. 14) the primary was drawn out to its most effective position, that is, the position which

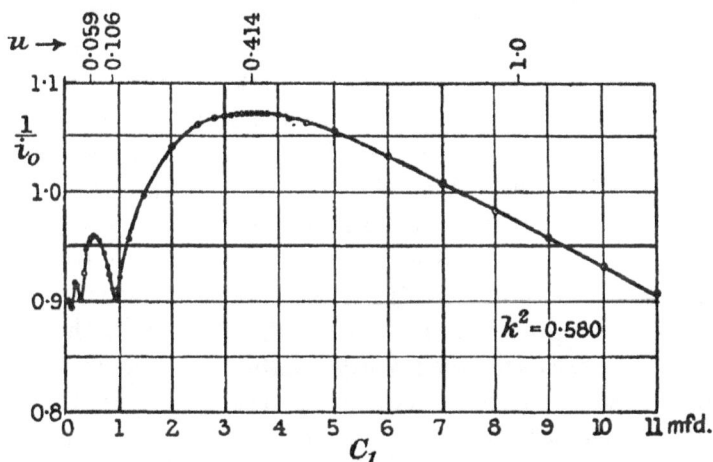

Fig. 14. Observed Variation of Secondary Potential with Primary Capacity at Coupling 0·58

caused the spark to appear at the smallest primary current. In determining this position, the primary capacity was, of course, adjusted to its optimum value for each of the positions compared with it. In the most effective arrangement the primary coil was more than 1 ft. from its usual symmetrical position. The coupling was found to be 0·580, practically the first of the special values of Table I, page 32.

The curve of Fig. 14 should be compared with the full line curve of Fig. 9, the theoretical curve for an ideal coil in which the coupling is $k^2 = 0·571$, the first of the values which allow of unit efficiency of conversion. It will be seen that the two curves bear a strong resemblance to each other in general form, each consisting of a series of arches diminishing in height with diminishing u. The relative proportions of the successive

arches are substantially similar in the two curves, though there
are certain differences in detail. For example, the ratio of
the greatest maximum to the nearest minimum is, in Fig. 9,
1·26 as against 1·19 in the experimental curve of Fig. 14. It
should, however, be remarked in this connection that it is not
easy to determine the experimental minima very exactly. Near
their points of intersection the arched curves are very steep,
so that unless one can proceed by very small steps in the
variation of C_1 one may easily pass over the exact minimum
point, and the true minimum may be appreciably less than
the minimum actually observed. There is also another reason
why the ordinates at the minimum points are rather too great,
both in Fig. 14 and in the other curves about to be described.
It will be shown later that in the adjustments giving these
minimum points the curve of secondary potential (V_2, t) is of
the double-peaked variety corresponding to the frequency-
ratios 5, 9, . . . of the system. In these cases it is found that
the spark passes more easily in the second peak than in the
first, even though the potential is rather lower in the second
than in the first. In other words, if the two peaks are nearly
equal, the least sparking current generally causes the spark
to pass at the *second* peak. The least sparking current may,
therefore, fail in such cases to be a reliable indication of the
maximum secondary potential. This matter will be referred to
again in connection with certain other phenomena.

On the other hand, the maxima in the experimental capacity
potential curves can be determined with greater accuracy, and
closer agreement is found in the ratios of successive maxima
between the curves of Figs. 9 and 14. Taking the ratio of the
greatest to the second maximum in Fig. 9, we find the value
1·146; in the experimental curve of Fig. 14 the ratio is 1·116.
The ratio of the second to the third maximum is 1·056 in Fig. 9,
and 1·048 in Fig. 14.

The two curves also show fairly good agreement in regard
to the values of u at which the maxima and minima occur.
The greatest maximum occurs at $u = 0·429$ in the ideal curve
(Fig. 9), at 0·414 in the experimental curve ; the first minimum
at $u = 0·105$ in the ideal curve, the experimental value being
0·106.

In Figs. 15, 16, and 17 are shown the experimental curves obtained in three other positions of the primary coil, the values of the coupling in these positions being respectively 0·699, 0·767,

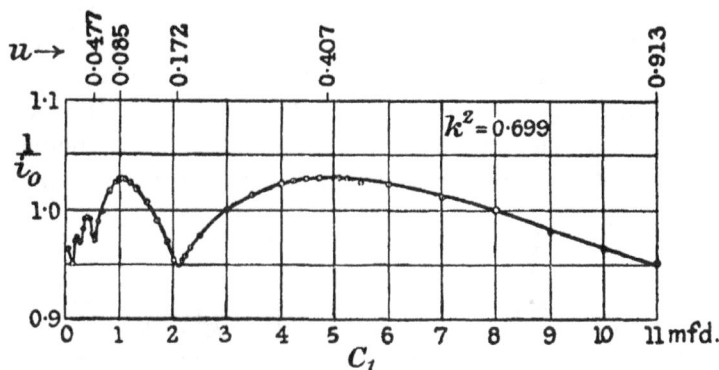

FIG. 15. OBSERVED VARIATION OF SECONDARY POTENTIAL WITH PRIMARY CAPACITY AT COUPLING 0·699

and 0·832. These curves should be compared with the theoretical curves of Figs. 10, 11, and 12, which are calculated for an ideal coil having nearly the same degrees of coupling. It

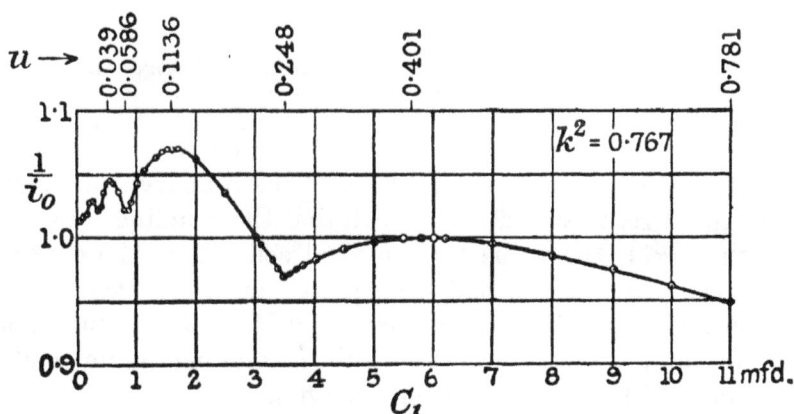

FIG. 16. OBSERVED VARIATION OF SECONDARY POTENTIAL WITH PRIMARY CAPACITY AT COUPLING 0·767

will be found on making the comparison that there is in these cases, as in the case just described, general agreement in form between the ideal and experimental curves, with similar points of difference. The most marked general difference between the

two sets of curves is that the percentage drop from the greatest
maximum to the neighbouring minima is somewhat less in the
experimental than in the ideal curves. The reasons for this
have been already explained. The percentage drop from the
greatest maximum to the next maximum (on the left) is also
rather greater in the ideal than in the experimental curves,
and this difference increases with the coupling. It appears,
therefore, that the effect of the resistances in reducing the
secondary potential is greater the greater the primary capacity,

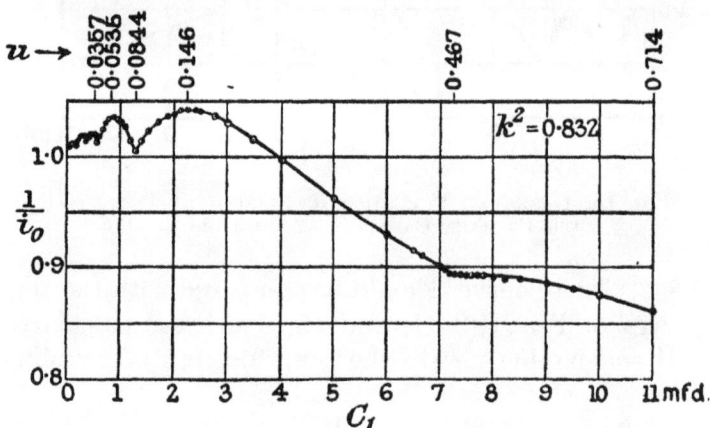

Fig. 17. Observed Variation of Secondary Potential
with Primary Capacity at Coupling 0·832

and especially so when the coils are closely coupled. One consequence of this is that the position of the primary coil which
gives equal greatest maxima is such that the coupling is rather
less than 0·71, the value for the ideal coil. If, in the experiment, the primary coil had been placed rather farther inside
the secondary, so that k^2 was raised to 0·71—a displacement of
about 5 mm. would have sufficed—the second maximum would
have been distinctly greater than the first.

Except in the case of Figs. 12 and 17, the values of u at
which the chief maxima and the points of intersection of the
arches occur, agree well in the ideal and experimental curves.
In Fig. 17, however ($k^2 = 0\cdot832$), these values of u are all
less than the corresponding values in Fig. 12 ($k^2 = 0\cdot835$).
Thus the greatest maximum occurs at $u = 0\cdot146$ in Fig. 17,
at 0·165 in Fig. 12, the neighbouring minimum at 0·0844 in

Fig. 17, at 0·0862 in Fig. 12. These differences cannot be attributed to a constant error in L_1 or in L_2C_2, but more probably arise from the general distortion—greater at the higher degrees of coupling—of the capacity-potential curves due to the resistances.

Use of the Curves for finding the Coupling. The general conclusion which we may draw from the experiments just described and the theoretical curves of Figs. 9–13, is that for each degree of coupling there is a capacity-potential curve of well-marked type which is not seriously modified by the resistances in a well-constructed modern coil of the usual form. This consideration suggests that it is possible to determine approximately the coupling of an induction coil by spark-length (or minimum sparking current) observation alone. If two or three of the principal arches of the capacity-potential curve are thus determined, a comparison with the curves of Figs. 9–13 will enable one to form an idea of the value of the coupling. A fairly accurate value may probably be arrived at by taking the ratio of the two greatest maxima, or the ratio of the values at C_1 at which these maxima occur, these features of the curve being apparently least subject to modification by the resistances. For example, if the ratio of the first to the second maximum is 1·13, and the ratio of the capacities at which they occur is about 7, the coupling is not far from $k^2 = 0·57$. Two equal greatest maxima at capacities in the ratio 4·8 indicate the value $k^2 = 0·7$, while equal greatest maxima with a capacity ratio of about 2·8 point to $k^2 = 0·87$.

As an example, the coil used in the experiments just described, with the primary coil in its usual position and with the secondary condenser, gives the curve shown in Fig. 17, indicating that the coupling is 0·832. When the secondary condenser was disconnected, the curve obtained was that shown in Fig. 18, the spark-length in this experiment being 9·6 mm. The proportions of the arches in Fig. 18 are approximately the same as those of Fig. 16, and indicate that the coupling is now about 0·767. The result shows that if the oscillating current in the secondary coil changes from one of uniform to one of non-uniform distribution (as it does when the secondary condenser is removed) the coupling is diminished.

The curve in Fig. 19 was obtained with the same arrangement but with a much greater spark-length, viz. 15 cm. between spherical electrodes.

The curve has much the same proportions as that of Fig. 18, showing that over a considerable range the coupling and the optimum primary capacity are practically independent of the current.

The present theory, as illustrated by the curves of Figs. 9 to 13, indicates that an induction coil has a certain optimum

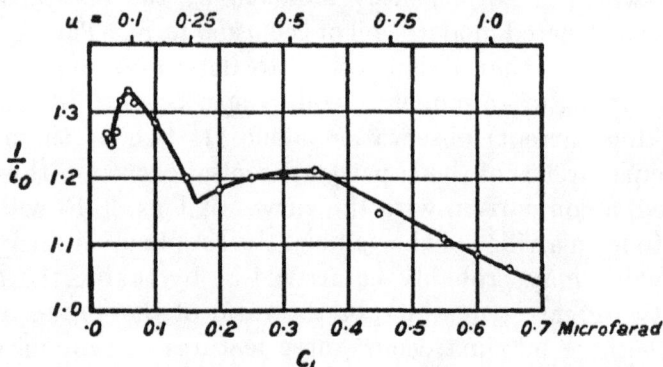

FIG. 18

When a condenser is connected with the secondary terminals the curve changes from this into that of Fig. 17, showing that the coupling is increased.

primary capacity, different from zero, even when the interruptions of the primary current are supposed to take place with perfect suddenness. Before 1914, several experimenters had examined the relation between primary capacity and spark-length, and all had come to the conclusion that as the capacity is diminished the spark-length rises to a single maximum from which it falls rapidly at small values of C_1. It is difficult to explain why no arches were observed at that time in the (C_1, V_{2m}) curves. It was not due to inferiority in the construction of coils used at that time, for the writer has obtained the arched curves with a very old coil. Probably it was due to some inferiority in the interrupters or the spark gaps used by those experimenters. The rapid fall in the spark-length at small values of C_1 was, in fact, attributed to failure of the interrupter, and it was generally believed that with a sufficiently

quick break a coil would work better without a primary condenser. This conclusion seemed to be supported by the late
Lord Rayleigh's rifle-bullet experiment. This experiment was
repeated by the present writer in 1914* with the coil used in
the experiment of Fig. 19, and in the same adjustment. The
object of the experiment was to determine whether the coil
gave a longer spark with a primary capacity of 0·05 mfd.
(nearly the theoretical optimum) than it did without any

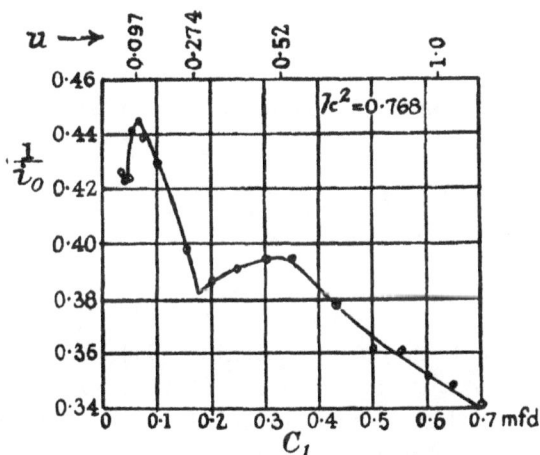

Fig. 19.   Curve Obtained with 15-cm. Sparks

condenser. The condenser could be connected to the ends of
a short piece of wire forming part of the primary circuit, and
the interruptions were effected by breaking the wire with a
bullet fired by a service rifle. The result of the experiments
was conclusively in favour of the presence of the condenser,
and this fact, together with the evidence afforded by the arched
(C_1, V_{2m}) curves described in the present chapter, shows clearly
that the action of induction coils is in the main correctly
described by the theory explained in these pages.

The Optimum Secondary Capacity. To consider another problem of a similar kind we will examine the effect on the maximum secondary potential of varying the secondary instead of
the primary capacity. At first sight, it might seem that the
value of V_{2m} would increase continually as the secondary

* *Phil. May.*, p. 582 (April, 1914).

capacity is diminished, and this would, of source, be the case if the maximum secondary *charge* were the same for all values of C_2. On the other hand, when the secondary capacity is varied, the frequency ratio changes, and at some values of C_2 the ratio will have those values which are most favourable to the production of high secondary potential, so that the (C_2, V_{2m}) curves might be expected to show maxima and minima similar to those of the (C_1, V_{2m}) curves, and C_2 might in consequence have a finite optimum value. The latter view is the correct one, and the proof is as follows. Employing the reciprocal ratio $m(= 1/u = L_2C_2/L_1C_1)$ we find by equations (42), (28), (29), for the maximum secondary potential the expression—

$$V_{2m} = \frac{L_{21}i_0}{\sqrt{(L_1C_1)}}\, M \sin \varphi \qquad . \qquad . \qquad . \qquad . \qquad . \qquad (44)$$

where $M^2 = \dfrac{1}{1 + m - 2\sqrt{\{m(1 - k^2)\}}}$

and the angle φ is given by equations (41). When C_2 alone is varied, the only variable factors in (44) are M and $\sin \varphi$. The quantity M is the same function of m that U is of u (see equation (27)), and has a maximum value $1/k$ when $m = 1 - k^2$. Also, the frequency ratio and the value of $\sin \varphi$ depend on the ratio u but are the same for any given value of this ratio and for its reciprocal, as is evident from the form of the expression (14). Consequently, the value of the product $M \sin \varphi$ for a given value of m is equal to that of $U \sin \varphi$ for the same value of u, and the value of m required to give the maximum of $M \sin \varphi$ is equal to that of u at the maximum of $U \sin \varphi$.

It follows that if we begin in both cases at an adjustment in which $L_2C_2 = L_1C_1$, the maximum secondary potential goes through a series of changes when we diminish the *secondary* capacity similar to those which it shows when the *primary* capacity is diminished. The $(u, U \sin \varphi)$ curves of Figs. 9 to 13 are applicable to the present problem if the abscissae are taken to represent m and the ordinates $M \sin \varphi$. So far, therefore, from the principal maximum secondary potential increasing indefinitely as C_2 is diminished, we find that, according to the present theory, it goes through a series of maxima and minima as C_2 is reduced, remaining finite when $C_2 = 0$.

In the adjustments giving maximum secondary potential when the secondary capacity is varied the values of m are necessarily less than unity, and are given, for various values of the coupling, by the numbers in the second column of Table II, page 38. In fact, the first five columns of Table II are applicable to these adjustments if the values of u are taken to represent m, and those of U to represent M.

The condition for maximum secondary potential, when the secondary capacity is the variable quantity, is, of course, quite different from the condition for maximum electrostatic *energy* in the secondary circuit. The latter is identical with the condition for maximum conversion efficiency, the adjustments for which have already been discussed. In the adjustments here under consideration, in which the secondary capacity alone is varied so as to give a maximum potential, the conversion efficiency has much smaller values. From equation (44) it is easily seen that the conversion efficiency, i.e.

$$\frac{L_{12}}{L_{21}} \cdot \frac{C_2 V_{2m}{}^2}{L_1 i_0{}^2},$$

is equal to $mk^2M^2 \sin^2\varphi$. The values of this quantity in the present adjustments are found by multiplying together the numbers in the second and sixth columns of Table II.

The general correctness of the above theoretical conclusions concerning the effect of varying C_2 may be verified by experiment. For this purpose it is advisable to employ a primary condenser of large capacity, say, 20 or 30 mfd., and to connect with the secondary terminals a condenser of variable capacity. If k^2 is about 0·57 it will be found that, on C_2 being reduced from its greatest value, the spark-length increases up to a certain point beyond which, with further reduction of C_2, the spark-length *diminishes*, showing that there is a certain optimum value of the secondary capacity. The diminution of C_2 being continued a minimum spark-length will be found, followed by a second maximum (smaller than the first) in accordance with the curve of Fig. 9, which, as explained above, should represent the relation between m and $M \sin \varphi$, as well as that between u and $U \sin \varphi$. If k^2 is approximately 0.71, it will be found that there are two nearly equal maxima of spark-length, occurring at different values of C_2 (compare Fig. 10).

The experiment is somewhat complicated by the fact that unless the capacity of the secondary condenser is very great in comparison with that of the secondary *coil*, the variation of C_2 causes a change in the distribution of the current in the secondary coil, and thus also gives rise to a change in its self-inductance. It is, in fact, usually impossible to vary C_2 without at the same time varying L_2. There is also evidence, as we have seen, that the coefficient of coupling of the primary and secondary coils varies with the nature of the distribution of the current in the secondary, and therefore also with the value of the capacity connected with it. For these reasons it is not so easy to determine the effect on the secondary potential of varying C_2 alone, and thus to obtain by experiment the $(m, M \sin \varphi)$ curve for a given value of k^2, as it is to examine the effect of varying the *primary* capacity, which is usually so large in comparison with that of its associated coil that the current in the latter may be regarded as always uniformly distributed along its length.

With regard to the secondary *charge*, the maximum value of this is given by

$$C_2 V_{2m} = \frac{L_{21} i_0 \sqrt{L_1 C_1}}{L_2} m\, M \sin \varphi \qquad . \qquad . \qquad . \quad (45)$$

On the assumption that C_2 alone is varied, k^2 having any constant value likely to be found in an induction coil, the function mM, i.e. $m/\sqrt{[1 + m - 2\sqrt{\{m(1 - k^2)\}}]}$ has no maximum or minimum value at any finite value of m; it increases continually as m increases from zero to infinity. Consequently, there is no finite "optimum" value of C_2 in relation to the secondary charge. The charge accumulated in the secondary condenser at the moment of maximum potential is greater the greater the capacity of this condenser.

Variation of the Coupling. We will now consider the manner in which the maximum secondary potential of a coil varies with the coupling of its circuits, the primary capacity at each value of k^2 being supposed to have its optimum value. The results indicated by the theory in this problem, for a coil of negligible resistances, are contained in Table II, page 38, but the interpretation of them depends upon the manner in which the variation of coupling is effected. In the first place, we will represent in a diagram the efficiency of conversion (for optimum adjustments) at different values of k^2. This curve is shown in Fig. 20, the abscissa being the coupling k^2, and the ordinate representing the efficiency of conversion, some values of which are given in the sixth column of Table II. The curve shows a maximum efficiency of unity at each of the values

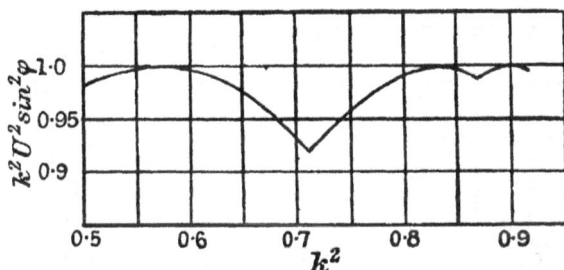

FIG. 20. RATIO OF SECONDARY ELECTROSTATIC TO PRIMARY MAGNETIC ENERGY, I.E. EFFICIENCY OF CONVERSION, AT DIFFERENT DEGREES OF COUPLING (CALCULATED)

of k^2, 0·571, 0·835, 0·902, which are the first three values given in Table I, page 32. It shows also minimum values of the efficiency at those values of k^2 (viz. 0·71, 0·87) at which the two principal arches in the capacity-potential curves are of equal height, that is, at which there are two optimum primary capacities (see Fig. 10).

Let us suppose that the secondary terminals of a coil are connected with a condenser of capacity large in comparison with that of the secondary coil, and that the variation of coupling is effected by drawing out the primary coil to various distances from its symmetrical position within the secondary, the optimum primary capacity, the least sparking current (for a constant gap), and the coupling, being determined for each position of the primary coil. In this experiment the initial electrokinetic energy is proportional to the square of the primary current, since the primary self-inductance is approximately constant. Also the maximum electrostatic energy in the secondary is constant, since the maximum secondary potential is constant, and the secondary capacity, which consists mainly of the capacity of the condenser, is not appreciably varied by changing the position of the primary coil. Consequently, the conversion efficiency is inversely proportional to the square of the primary current i_o.

The curve in Fig. 21 shows the results of an experiment made in this way, the abscissa representing k^2 and the ordinate $1/i_0^2$. This curve shows maxima and minima of efficiency at practically the same values of k^2 as those indicated in Fig. 20, and the ratio of the efficiency at $k^2 = 0.58$ to that at $k^2 = 0.7$ in the experimental curve is the same as the ratio of the maximum to the minimum in Fig. 20. The chief difference between the curves is that in the theoretical curve (Fig. 20) the efficiency at $k^2 = 0.835$ is equal to that at 0.571, while in the experimental curve it is appreciably less.

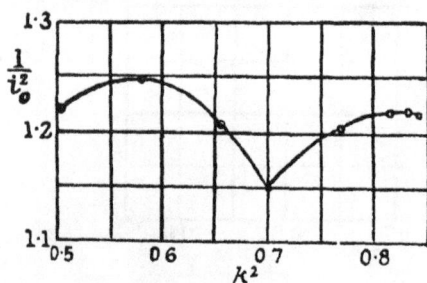

FIG. 21. OBSERVED EFFICIENCY OF CONVERSION AT DIFFERENT DEGREES OF COUPLING

The difference is too great to be accounted for by the small error arising from the variation of secondary capacity due to the displacement of the primary coil, and it must be regarded as showing that the effect of the core losses is greater at the higher degrees of coupling. The experimental value of the efficiency of conversion is, of course, much smaller than that indicated in Fig. 20 which refers to a coil devoid of resistances. In an actual coil, the efficiency of conversion may be over 60 per cent, but it can be much lower if the primary capacity is not adjusted to its optimum value.

With regard to the variation of the maximum secondary potential in the changes of adjustment just described, the value of V_{2m} for a given primary current is proportional to $1/i_0$. It is, therefore, proportional to the square root of the ordinate in Fig. 21, and has maxima and minima at those degrees of coupling which give the highest and lowest efficiency of conversion.

Series Inductance in the Primary Circuit. Another way of varying the coupling is to connect in the primary circuit coils having self-inductance which do not act inductively on the secondary. If the primary current at break is given, the increase of the total self-inductance of the primary circuit causes a proportional increase in the magnetic energy $\frac{1}{2}L_1i_0^2$ supplied to the system, and if the process is carried to a certain stage there will be no diminution—there will generally be an increase —of conversion efficiency, so that a considerable improvement in spark-length is to be expected.

The manner in which the maximum secondary potential changes when the primary self-inductance is increased is shown —for the ideal case of negligible resistances—by the values of $U \sin \varphi$ given in Table II, p. 38, since the factors L_{21} and L_2C_2 in the expression (42), p. 37, for the principal maximum secondary potential—i.e. $V_{2m} = L_{21}i_0U \sin \varphi / \sqrt{L_2C_2}$—are not altered by connecting external series inductance in the primary circuit. In Fig. 22 these values of $U \sin \varphi$, which correspond, of course, in each case to the optimum primary capacity, are plotted as a curve, the abscissa of which represents $1/k^2$, i.e. $L_1L_2/L_{12}L_{21}$, this quantity being proportional to the total self-inductance of the primary circuit. The full-line curve, the ordinate of which is proportional to the greatest maximum secondary potential at each stage, consists of three portions, A, B, C, corresponding to adjustments in which the frequency-ratio n_2/n_1 is near 3, 7, 11 respectively. The points at which these sections of the curve meet represent those cases in which there are two equal greatest maxima of V_2. The broken-line continuations of the curves at these points correspond to secondary maxima in the $(u, U \sin \varphi)$ curves. At each of these points of intersection also the efficiency of conversion is a minimum for "optimum" adjustments.

At the points marked a, b, c on the curve, the frequency-ratio has the exact values 3, 7, 11, and the conversion efficiency in these adjustments is a maximum. The efficiency curve is also shown in Fig. 22, by the line DEF, the ordinate of which represents $k^2U^2 \sin^2 \varphi$.

Of the adjustments which give the maximum efficiency of conversion, the one corresponding to the point a clearly gives

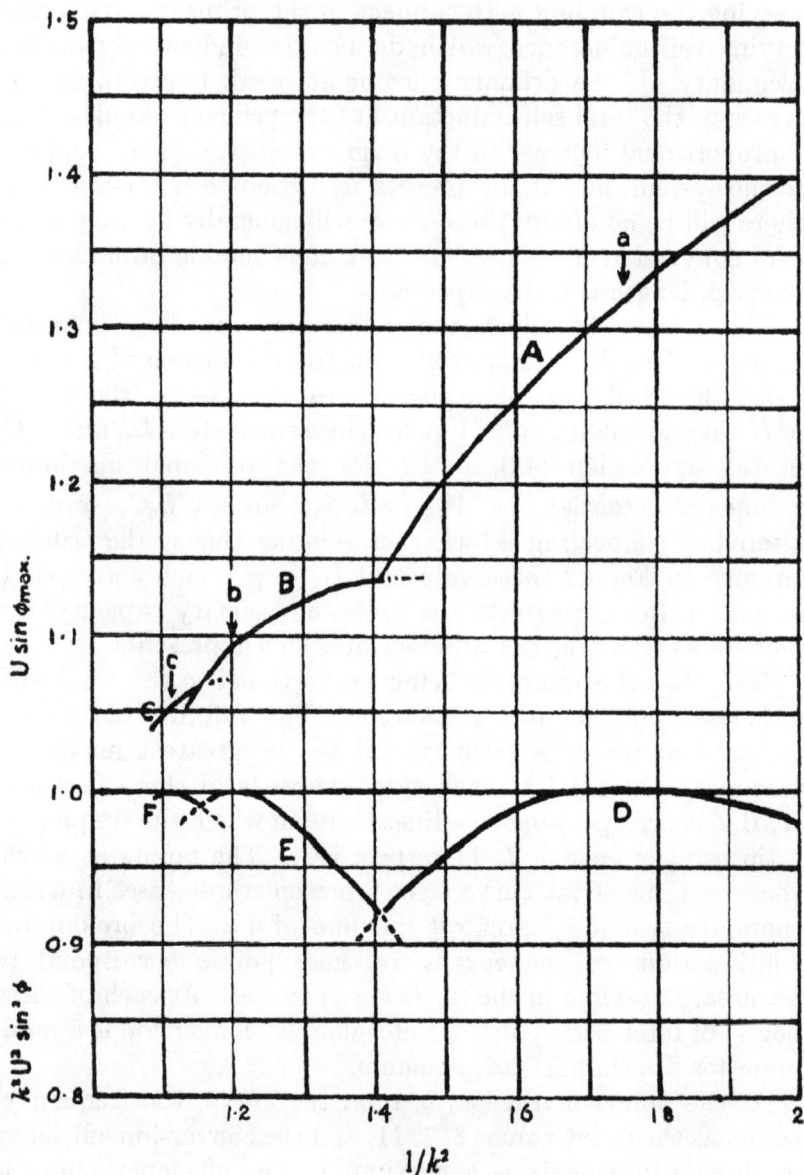

Fig. 22

Curve *CBA* shows the increase of maximum secondary potential accompanying the addition of series self-inductance to the primary circuit of an induction coil. The abscissa $1/k^2$ is proportional to the total self-inductance of the primary circuit. The ordinate represents the greatest maximum of $U \sin \varphi$ at each stage. Curve *FED* shows the variation of efficiency of conversion.

the highest secondary potential for a given primary current. In these adjustments the maximum potential is, in fact, given by

$$V_{2m} = i_0 \sqrt{\frac{L_{21}}{L_{12}}} \sqrt{\frac{L_1}{C_2}}$$

(see (33), page 31), so that the potential is proportional to the square root of the total primary self-inductance. The secondary potential at the point a ($k^2 = 0.571$, $u = 0.429$) is about 21 per cent higher than the value at the point b ($k^2 = 0.835$, $u = 0.165$). When the coupling is varied by adding series inductance to the primary circuit the first of the unit-efficiency adjustments (frequency-ratio 3) is therefore more effective than the others, in the sense that it allows a higher secondary potential to be developed at the interruption of a given primary current. If the primary inductance is increased beyond the point a, so that k^2 becomes less than 0.571, the secondary potential continues to increase, but with diminishing efficiency of conversion.

In the curves of Fig. 23 are shown the results of a series of measurements in which the coupling was varied by adding self-inductance to the primary circuit. The abscissa in this diagram represents $1/k^2$ which, as already explained, is proportional to the total self-inductance of the primary circuit. In the upper curve the ordinate is $1/i_0$, the reciprocal of the least sparking current for a constant gap, which is proportional to the maximum secondary potential for unit primary current. In the lower curve, the ordinate is k^2/i_0^2 which is inversely proportional to $L_1 i_0^2$, and is therefore—since the maximum secondary electrostatic energy is constant throughout the series —proportional to the efficiency of conversion.

The general similarity between the curves of Fig. 22 and those of Fig. 23 is obvious, and the numerical agreement between them is also in some respects very close. Thus, the values of $1/k^2$ at which the maximum and the minimum of efficiency occur (lower curves) are much the same in the two diagrams, and the increase of maximum secondary potential per unit primary current for a given increase of primary self-inductance (upper curves) is in the same proportion in an actual coil as it would be if the resistances were entirely negligible.

It may be noticed that the upper curves in Figs. 22 and 23 do not differ greatly from straight lines. It may, therefore, be stated as an approximate rule, applicable over a wide range,

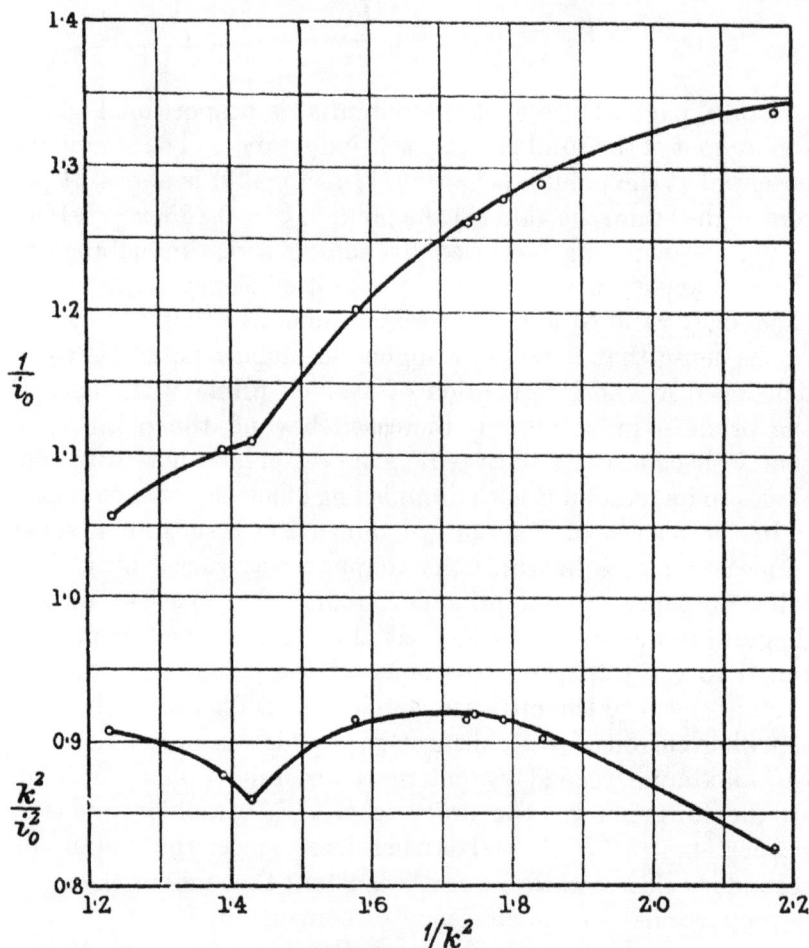

FIG. 23

The upper curve shows the variation of secondary potential, the lower curve that of the efficiency of conversion, with total primary self-inductance.

that if the coupling is reduced by adding series inductance to the primary circuit, the primary capacity being always adjusted to its optimum value, the increase of the maximum secondary potential for a given primary current is proportional to the increase of the self-inductance of the primary circuit.

The Effect of Partly Closing the Core. The variation of coupling caused by partly closing the iron core has been referred to in Chapter I. The present theory is applicable to such variation provided the core is not so nearly closed that the magnetizing current and the magnetic flux deviate too far from the approximate proportionality assumed. The effect on the secondary potential may be seen from the expression $L_{21}i_0\,U\sin\varphi/\sqrt{(L_2 C_2)}$ for this quantity. The factor $L_{21}/\sqrt{L_2}$ is increased by the reduction of magnetic reluctance of the core, since L_{21} and L_2 are inversely proportional to the reluctance. The factor $U\sin\varphi$ is slightly diminished owing to the increase in the coupling (see Table II, page 38, or Fig. 22, page 60). As to the secondary capacity C_2, there seems to be no reason for supposing that this quantity is sensibly altered. The principal effect is thus the increase of $L_{21}/\sqrt{L_2}$, and the theory therefore indicates that a higher secondary potential should be obtained for a given primary current with the nearly closed than with the usual straight core. On the other hand, if the coupling is very close, the factor $U\sin\varphi$ cannot be large (see Table II), and a considerable improvement in spark-length for a given current can be effected without loss of efficiency by connecting suitable series inductance in the primary circuit so as to bring the system into the first of the adjustments of Table I, page 32.

Reduction of Number of Secondary Turns. Some coils are made with a comparatively small number of secondary windings, and we will, therefore, consider here briefly the effect to be expected as a result of a diminution of this number. For this purpose, we may suppose the system to be in one of the adjustments of Table I, so that the maximum secondary potential is, by (33), page 31,

$$V_{2m} = i_0\sqrt{\frac{L_{21}}{L_{12}}}\sqrt{\frac{L_1}{C_2}}.$$

A reduction of the number of secondary turns will generally involve a diminution of the secondary capacity C_2, which is the only variable quantity in the expression for V_{2m}. Consequently, if the system can be kept in the same adjustment we should expect a diminution of the number of secondary turns to result in an increased secondary potential for a given primary current.

On the other hand, the reduction of the secondary turns involves a more than proportional diminution of the product L_2C_2, and therefore also of the primary capacity required to keep the system in adjustment. Now, one of the factors upon which the effective working of an interrupter depends is the initial rate of rise of the primary potential V_1, and this, by the first of equations (20), is equal to i_0/C_1. The effectiveness of an interrupter in breaking strong currents is, therefore, reduced by diminishing the capacity associated with it, and this fact sets a limit to the advantage in secondary potential which may be gained by reducing the number of secondary turns.

Change of Dimensions of a Coil. Let us suppose that all the linear dimensions of an induction coil, including the length and diameter of the core and of the windings, the length and breadth of the plates of the condenser, and the thickness of the insulation in all parts, are increased in the same ratio. We shall assume that the coil is initially in one of the unit-efficiency adjustments (see Table I, page 32), and shall neglect damping, so that the maximum secondary potential is given by

$$V_{2m} = i_0 \sqrt{\frac{L_1}{C_2}} \sqrt{\frac{L_{21}}{L_{12}}}.$$

By the change of size the inductances and capacities of the system are all increased in proportion to the linear dimensions, consequently the coupling and the ratio L_1C_1/L_2C_2 are unaltered and the above equation still holds. Since the second and third factors in the expression for V_{2m} are unchanged, it follows that any improvement in the secondary potential can only arise from the use of a stronger primary current i_0. On the view that the arcing tendency at the interrupter depends upon the ratio i_0/C_1 we should suppose that the primary current is increased in proportion to the capacity of the condenser; if the original current was the greatest that could be efficiently interrupted, the new current will then also possess this property. Further, in accordance with the principle of similarity as stated by Lord Kelvin for electromagnetic systems, if the current is increased in proportion to the linear dimensions the magnetic force and the flux-density in the core are unaltered, so that if the original current was such as to give the inductances their greatest values, the new inductances will also be maxima. We shall therefore assume that the primary current is increased in proportion to the length of the coil, and it follows that the maximum secondary potential is also increased in the same ratio.

The question arises here whether the change of dimensions would be accompanied by such increase of the effective resistances of the coil as would greatly increase the damping of the oscillations and so reduce the maximum secondary potential. The general nature of the effect of change of dimensions on the effective resistances may be seen by considering a single coil with laminated iron core, having self-inductance L and effective resistance R, the terminals of which are connected to a condenser of such capacity that the period of oscillation is T. The decrement of the oscillation of the coil, reckoned for a quarter of a period (the interval with which we are concerned in induction coil problems), is $RT/8L$.

It is known that by far the greater part of the effective resistance of the coil arises from losses in the core due to eddy currents and hysteresis. With regard to the eddy current loss, the increase of effective resistance of the coil due to this cause depends greatly on the thickness of the core laminae, and it is inversely proportional to the square of the period of oscillation. It can be shown that if the thickness of the core sheets is not altered, the change of dimensions of the coil has no effect on the ratio R/L so far as the eddy current loss is concerned, but since the period T is increased in proportion to the linear dimensions, the decrement is also increased in this ratio. As to the hysteresis loss, it can be shown that the increase of effective resistance due to this cause in the change of dimensions has no effect on the decrement. On the whole, it is to be expected that there will be some increase of effective resistance (due to eddy currents) when the dimensions of the coil are enlarged, even on the supposition that the thickness of the core sheets is not altered.

We have supposed, however, that the number of turns in the primary and the secondary coil is not altered when the dimensions are changed, but that the diameter of each wire is increased in proportion to the length or the diameter of the coil. The area of the cross-section of each wire is, therefore, increased in proportion to the square of the linear dimensions, and is unnecessarily great for currents which are only proportional to their first power. It is probable that the effect of the increased effective resistances of the enlarged coil could be more than compensated by reducing the thickness of the primary and secondary wires and increasing their numbers of turns, thus increasing L_1 without increasing C_2 and without altering the adjustment of the system. There seems to be no good reason for doubting, therefore, that the maximum secondary potential obtained with a coil is proportional to its linear dimensions.

A certain coil examined by the writer, when used with a series inductance in the primary circuit, gave 239,000 volts at the interruption of a current of 4 amp. It appears reasonable to expect that a coil four times as large, with series coil and condenser increased in the same ratio, would give a million volts with a primary current of about 16 amp.

CHAPTER III

OSCILLOGRAPHS. WAVE FORMS

In the experimental study of induction coils and transformers, it is very desirable to have some convenient means of observing wave forms of secondary potential, i.e. the (V_2, t) curves, which show the manner in which the secondary potential varies with the time. An examination of these curves gives us information as to the component oscillations present and the way in which the potential grows to its maximum value, and affords a method independent of spark-length observation for comparing the maximum potentials in different cases.

It need scarcely be said that the instrument used in obtaining the curves should not itself take any current, i.e. it should be an electrostatic instrument, and should have very small capacity, so that its addition to the secondary circuit does not too greatly alter the period of oscillation of this circuit or the distribution of current along its length.

At the present time cathode ray oscillographs, in which a wave form is traced by a narrow pencil of cathode rays focused on a fluorescent screen, form a familiar part of the equipment of most Physics laboratories. These instruments are, however, usually much too sensitive to be suitable for direct connection to the secondary terminals of an induction coil, and they require a reduction of the potential to a suitable value by the use of condensers or high resistances, the presence of which is apt to modify considerably the conditions of the circuit. For photographic purposes, also, the cathode ray oscillograph is inferior to those instruments in which the wave form is shown by an intense pencil of light reflected from a small moving mirror and sharply focused on the plate. The disadvantage of mirror oscillographs arises from the limited natural frequency of vibration of the moving parts (when undamped), which makes these instruments quite unsuitable for the delineation of high-frequency wave forms. For photographing the wave forms of oscillations of moderate frequency mirror oscillographs are, however, more suitable and convenient than instruments of the cathode ray type.

Electrostatic Oscillograph. The following is a description of a mirror electrostatic oscillograph designed by the writer* for this purpose. It has very small capacity, can be connected directly to the secondary terminals of a coil or transformer, and may be used over a wide range of voltage, viz. 2,000 to over 200,000 volts; the upper limit (if such exists) has not been ascertained.

A piece of phosphor-bronze strip, *S* (Fig. 24), is soldered at one end to a terminal on an ebonite pillar, *P*, and at the

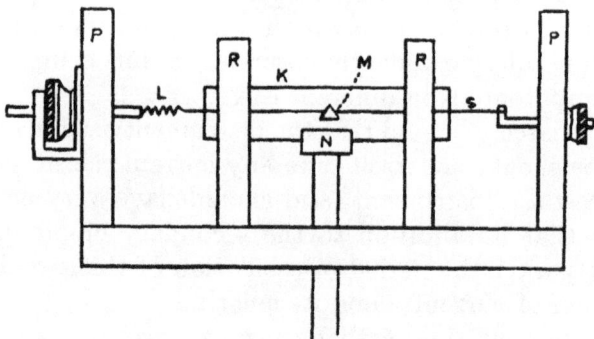

Fig. 24. Diagram of Electrostatic Oscillograph

P P = Ebonite pillars.	*L* = Spring.
R R = Glass rods.	*K* = Ebonite sheath containing
S = Strip.	attracting plate.
M = Mirror.	*N* = Adjustable platform.

other to a small spiral spring, *L*, attached to a screw. A second ebonite pillar supports the screw which can be drawn through by a nut, thus allowing the tension of the strip to be varied. The strip rests horizontally against two vertical glass rods, *R*, about 1·5 cm. apart. To the middle of the strip is attached a small mirror, *M*, of very thin silvered glass, rectangular or triangular in form, and 1 or 2 sq. mm. in area. In front of the strip, and connected to the terminal by a thin wire, is a thin plate of copper (not shown in the figure) bent so that its edge faces the strip and is less than 1 mm. from it. A gap in this plate allows a beam of light to pass to and from the mirror. Behind the strip is another thin plate of copper imbedded in a sheath of ebonite, *K*, and also with its edge facing

* *Phil. Mag.*, p. 238 (Aug., 1907).

the strip. The whole is mounted on an ebonite support, inverted, and placed in an ebonite vessel provided with a small window and filled with a transparent, insulating oil of suitable viscosity and of fairly high dielectric strength. It is also desirable that the oil should have a high specific inductive capacity. Heavy paraffin oil or castor oil, the latter diluted if necessary with a thinner oil, may be used as the damping liquid.

A small platform, N, of ebonite tipped with cork can be raised by a screw until it comes into contact with the lower edge of the mirror. A horizontal adjustment of the platform then allows the mirror to be tilted so that the reflected ray can be made horizontal, or can be given any desired small elevation.

The phosphor-bronze strip is thus between two plates, to one of which it is connected. If the plates are charged to a difference of potential, the strip is repelled by one plate and attracted by the other. The mirror then, since one of its edges is fixed, is deflected through a small angle proportional to the square of the difference of potential of the plates, as in the idiostatic use of the quadrant electrometer. A beam of light, proceeding from a small circular aperture illuminated by an arc lamp and condenser, passes through a convex lens and is reflected by the small mirror on to a rotating concave mirror, driven by a motor. It is finally focused so that an image of the aperture is formed on a plate of ground glass or on a photographic plate.

A tuning-fork carrying a small mirror is mounted vertically in front of and close to the window of the oil vessel. A part of the beam of light falls upon this mirror, whence it is reflected to the rotating mirror and focused on the plate. When the concave mirror rotates, two horizontal lines are thus traced by the spots on the ground-glass plate. By adjusting the small platform one of these traces can be brought to a short distance from the other. At a certain point of the rotation of the mirror the tuning-fork is struck by a hammer worked by a lever, after the manner of a pianoforte-key action. Another lever may be used for working the interrupter.

The arc lamp is enclosed in a light-tight case and the whole apparatus is set up in a dark room.

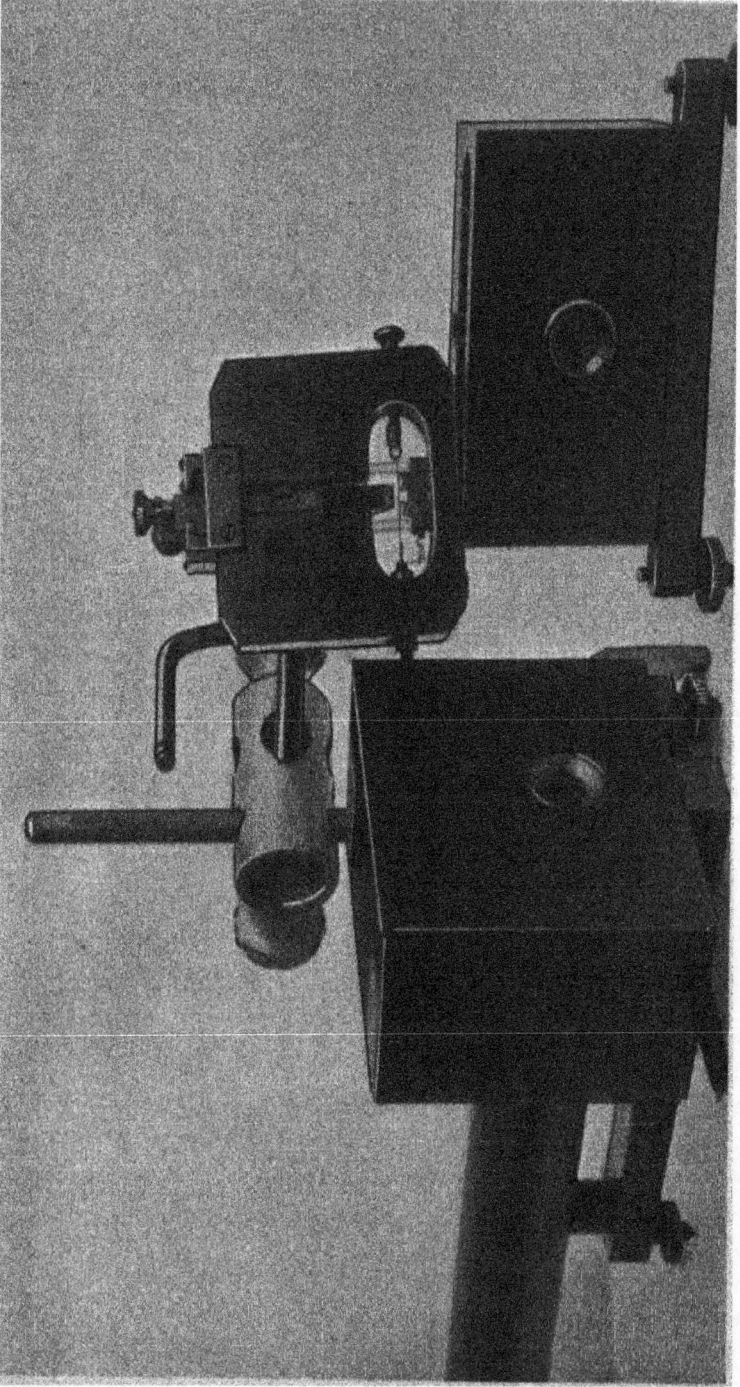

FIG. 25. ELECTROSTATIC OSCILLOGRAPH WITH OIL BATHS FOR TWO RANGES OF VOLTAGE

The same moving system may be used for all ranges of potential, but different ebonite sheaths with the attracting plates imbedded in them are required. These may be screwed in at the back of the oil bath, or a number of oil baths may

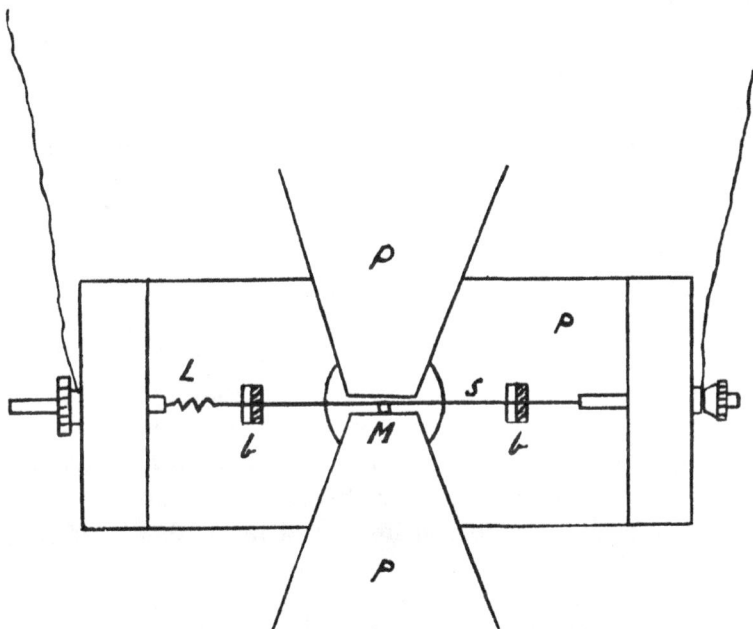

FIG. 26. DIAGRAM OF CURRENT OSCILLOGRAPH

P = Ebonite plate.	*M* = Mirror.
b b = Glass "bridges."	*L* = Spring.
S = Strip.	*p p* = Magnet pole pieces.

be used with the sheaths and attracting plates fixed in them. For the lowest potentials, no ebonite sheath is required, the bare attracting plate being brought close up to the strip. For 200,000 volts the strip and the attracting plate must be separated by 1 cm. of ebonite.

Fig. 25 shows an electrostatic oscillograph of this kind made for the writer by the Cox-Cavendish Electrical Company.

It has already been stated that the deflection of the mirror is proportional to the *square* of the potential difference applied to the terminals of the instrument. This is in some respects a disadvantage—the curves would in some cases be easier of interpretation if the deflexion were simply proportional to the potential—but any such disadvantage is largely outweighed by the greater simplicity, both in the construction and in the use of the instrument, arising from the idiostatic method of connection. It is also an advantage of this method, when the photographed curve is to be used for the measurement of a period of oscillation, that the minimum, i.e. zero, points of the curve, which frequently form excellent marks for measurement, all lie in or near the zero line. (See, for example, Fig. 28 on page 74.)

Current Oscillograph. A somewhat similar method may be employed in the construction of a simple form of instrument for showing the wave form of the current. For this purpose the strip is mounted on an ebonite plate P (Fig. 26) which has a central aperture to admit the light, and which carries also two short projecting glass "bridges" against which the strip rests. The plate P is supported by a brass rod (see Fig. 27) which can be moved vertically and horizontally so that the strip and mirror can be brought within the narrow gap between the polepieces of a permanent magnet. One of the polepieces is tipped with a very thin sheet of cork for holding one edge of the mirror. The whole is immersed in an ebonite oil-bath, and the instrument is suitable for use in a high- or a low-tension circuit. Fig. 27 shows the parts of a current oscillograph made in this way.* In this instrument, the gap between the pole faces is 1·5 mm. wide, and the instrument is suitable for currents up to 0·2 amp. in the strip, shunts being used for stronger currents.

Wave Forms of Secondary Potential. In all the oscillograms shown in this book the movement is from right to left, and the lower curve is the tuning-fork curve each wave of which represents 1/768 second.

The following series of photographs, Figs. 28 to 37, shows a number of oscillograms obtained with the electrostatic oscillograph which was connected directly with the secondary terminals of an induction coil. In each case, the terminals were insulated, so that no discharge passed between them. The

---

* The writer is much indebted to Messrs. Joseph Lucas & Co., who (through Dr. J. D. Morgan) kindly presented the horseshoe magnet used in this instrument, and to Messrs. Hadfields, Ltd., who generously gave the soft iron for the polepieces. Both the magnet and the iron have proved to be admirably suited for their purposes.

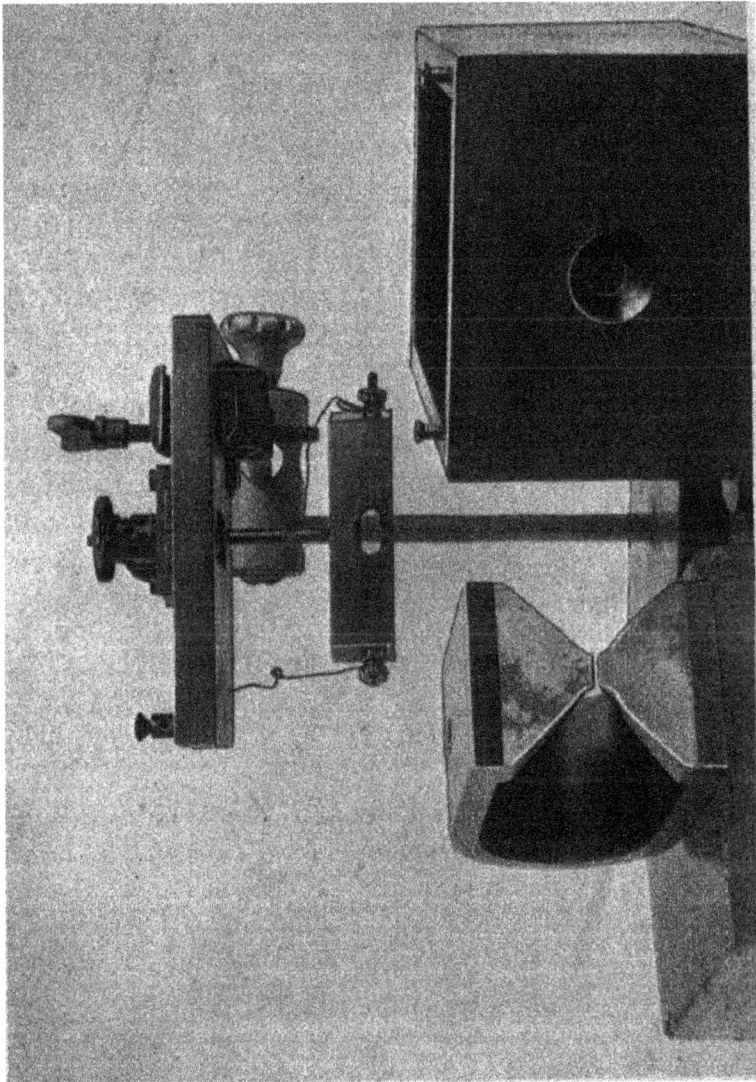

Fig. 27. Current Oscillograph with Magnet and Oil Bath

curves indicate the wave of potential set up in the secondary at "break."*

In Fig. 28, only one oscillation (frequency about 200) is in evidence. The constants of the circuits in this experiment were

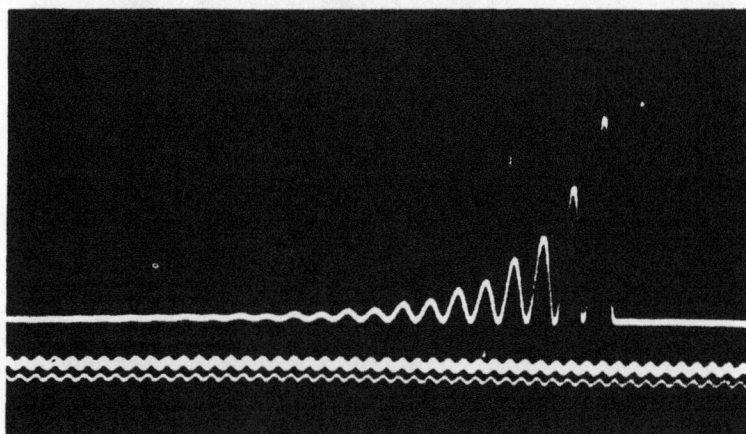

FIG. 28. POTENTIAL OSCILLOGRAM. SINGLE OSCILLATION
The deflection is proportional to the square of the potential.

such that the frequency ratio was 24, and the more rapid component, having an initial amplitude 24 times as small as that of the oscillation shown and being also more strongly damped, does not appear in the photograph.

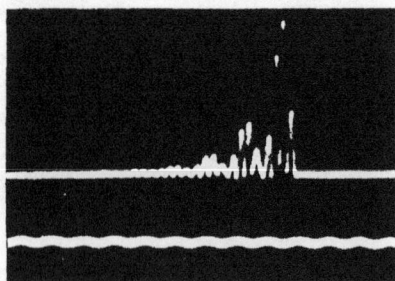

FIG. 29. DOUBLE OSCILLATION. FREQUENCY RATIO 3/2

In the experiment in which the photograph shown in Fig. 29 was taken the usual iron-core primary was replaced by a coil without an iron core in order to increase the frequency of the slower component and, by diminishing the coupling, to reduce

* See also *Journ. Rönt. Soc.*, Vol. XVI, April, 1920.

the frequency ratio. The curve evidently represents a double oscillation, the frequency ratio is about 1·5, and the actual frequencies are approximately 1,400 and 2,100. The curve shows the periodic variations of amplitude—the "beats" of

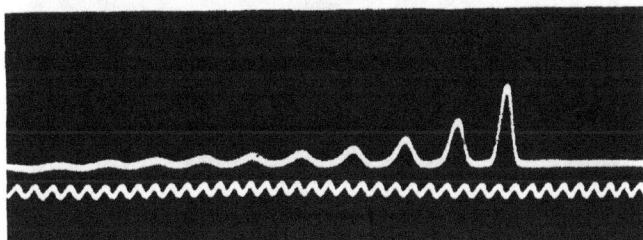

FIG. 30. DOUBLE OSCILLATION. FREQUENCY RATIO 3

acoustics—which occur when two oscillations of rather small frequency ratio are superposed. In this case there are 700 "beats" per second. In this experiment the frequency of the more rapid component was rather too great to allow the wave of potential to be accurately shown by the oscillograph, some

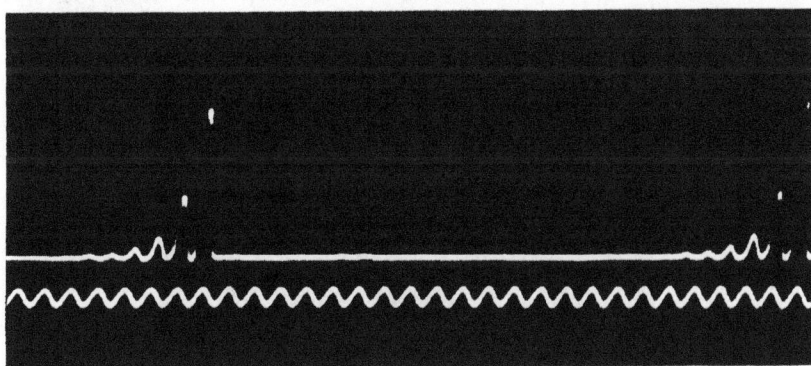

FIG. 31. TWO BREAKS AND A MAKE BY MOTOR INTERRUPTER.
MAXIMUM POTENTIAL 118,000 VOLTS

of the minimum points which ought to be in the zero line being considerably above it.

In Fig. 30 the frequency ratio is 3. The coupling in this experiment was reduced to the value 0·58 by moving the primary coil (iron cored) to a suitable distance along the axis of the secondary, and the capacity was then adjusted to its optimum value (see page 47). The system was, therefore, practically in the first adjustment of Table 1 (page 32). The oscillogram in Fig. 30 shows the peaked maxima and the

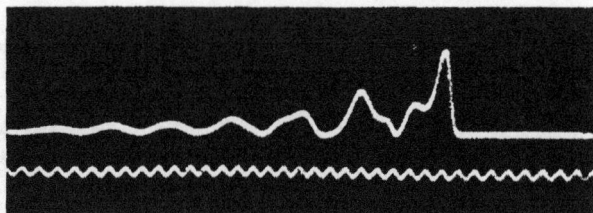

FIG. 32. FREQUENCY RATIO 3·75

flattened zeroes characteristic of the 3/1 frequency ratio (compare Fig. 6, page 30).

Fig. 31 shows another example of a 3/1 ratio curve, a motor mercury interrupter being used in this experiment. The curve shows the effect of two successive breaks, that of the comparatively very small secondary potential at "make" being just noticeable in the photograph at about three-quarters of the interval between the two breaks. In the experiment of Fig. 31,

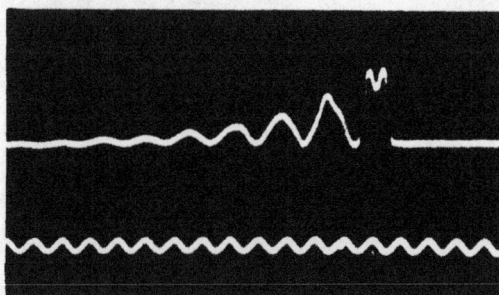

FIG. 33. FREQUENCY RATIO 5

the system was again nearly in its most efficient adjustment, the coupling in this case being reduced to the value 0·56 by the addition of series inductance to the primary circuit, and the primary capacity being the optimum (see Figs. 22, 23). The maximum potential was 118,000 volts, the mean primary current 0·3 amp., and the current at the moment of break rather over 2 amp.

In Fig. 32 the frequency ratio is 3·75; in Fig. 33 it is 5

(compare Fig. 6), the potential at each of the two equal peaks of
the curve in the latter case being 55,000 volts. In Fig. 34 the

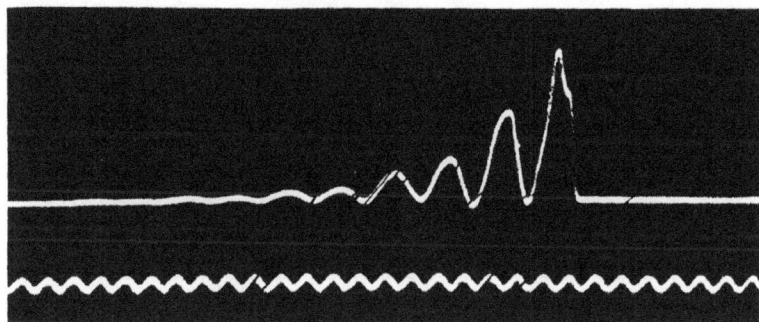

Fɪɢ. 34. FʀᴇQUᴇɴᴄʏ Rᴀᴛɪᴏ 6. Mᴀxɪᴍᴜᴍ Pᴏᴛᴇɴᴛɪᴀʟ 125,000 Vᴏʟᴛs

frequency ratio is nearly 6, and the maximum potential indicated is 125,000 volts.

Fig. 35 is the oscillogram for a case in which the frequency
ratio was 7, the second maximum of the more rapid oscillation
coinciding with the first maximum of the slower component

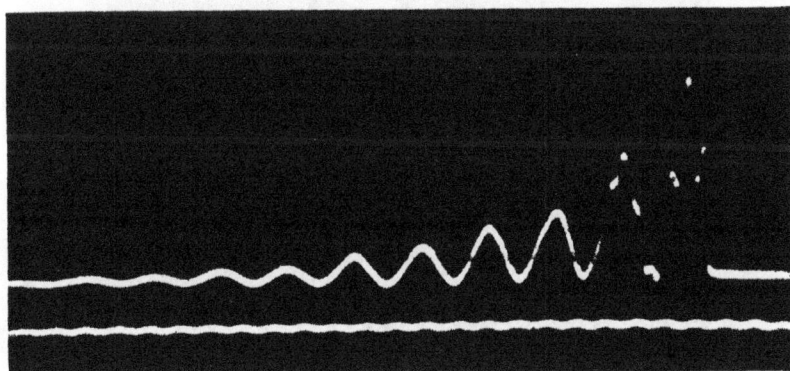

Fɪɢ. 35. FʀᴇQUᴇɴᴄʏ Rᴀᴛɪᴏ 7

(compare Fig. 6). In Fig. 36 the ratio is nearly 9, the third
minimum of the higher frequency component occurring very
slightly after the first peak of the main oscillation (see Fig. 6).
In this experiment the system was approximately in the adjustment corresponding to the second minimum (counting from
the right) in the capacity-potential curves of Fig. 10 (page 43)

and Fig. 15 (page 49), the coupling being 0·7 and the ratio L_1C_1/L_2C_2 being 0·046.

These examples show that the experimental wave form of the secondary potential of a coil is determined by the ratio

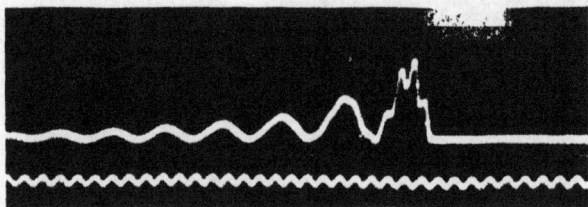

FIG. 36. FREQUENCY RATIO 9

of the frequencies of the two component oscillations in the manner indicated by the theory of the ideal coil (without resistances) explained in Chapter II. Some examples of the comparison of the experimental wave forms with the curves

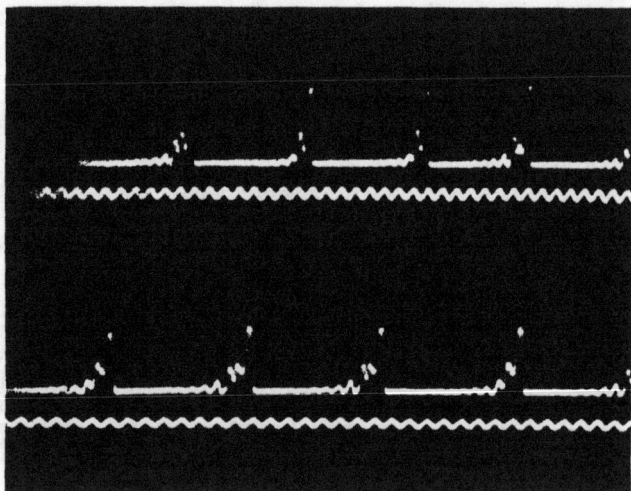

FIG. 37. SECONDARY POTENTIAL WAVE FORMS. WEHNELT INTERRUPTER

calculated from the constants (including the resistances) of the circuits will be given later in the present chapter.

Fig. 37 shows two wave forms of the secondary potential of a coil worked by a Wehnelt interrupter, the lower curve being taken when the secondary terminals were connected with an X-ray tube, the upper curve when the terminals were insulated. The tube was "hard" at the time and was not passing much current. The chief difference between the curves is in the value of the potential, which is rather lowered in the lower curve by the discharge. The primary current was the same for both curves. The form of these curves suggests that in the Wehnelt interrupter "make" occurs very soon after "break."

The High Tension Transformer. A high tension transformer may be worked with a battery and interrupter in the manner

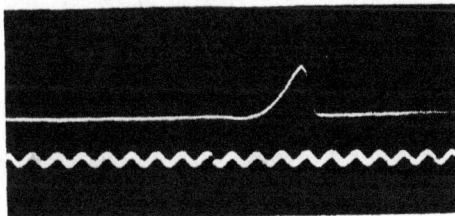

FIG. 38. CLOSED CORE TRANSFORMER WITH INTERRUPTER.
SECONDARY POTENTIAL

usual with induction coils, and an examination of a transformer used in this way yields valuable information as to the effect on the secondary potential of a coil of closing its iron core. The chief effects to be expected as a result of closing the core are increase of the coupling and of the core losses. Consequently, we should expect the oscillations of a transformer to be characterized by high frequency ratio and strong damping, both of which anticipations are borne out by experiment. Fig. 38 shows the wave of secondary potential of a closed-core transformer at the interruption of a primary current of 4·5 amp. supplied by a battery. The curve represents a single very strongly damped oscillation of maximum potential 96,800 volts, a performance which (owing to the damping) is much inferior to that of an ordinary open-core induction coil. In the experiment of Fig. 38 the primary capacity was 0·2 mfd., and the more rapid component oscillation was too small and too strongly damped to be shown in the photograph. The higher frequency component may be brought into evidence by increasing the primary capacity or by adding series inductance to the primary circuit so as to reduce the coupling.

Fig. 39 shows two photographs obtained when a small series inductance was included in the primary circuit of the transformer. In the lower curve the primary capacity was 0·3 mfd., in the

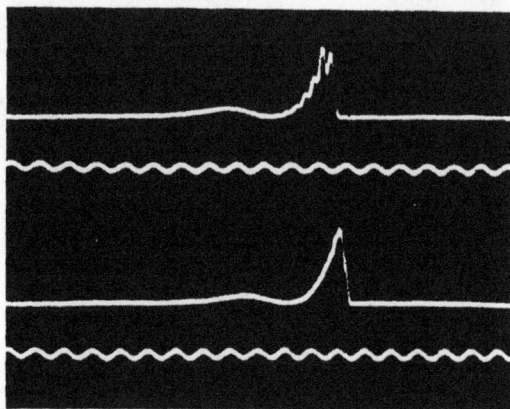

FIG. 39. SECONDARY OSCILLATIONS OF TRANSFORMER WITH SERIES INDUCTANCE AND CONDENSER IN PRIMARY

upper curve 1 mfd. In both curves the more rapid oscillation is plainly shown, having still, however, a frequency many times as great as that of the slower component. The primary current at break was the same in both curves of Fig. 39 (4·8 amp.),

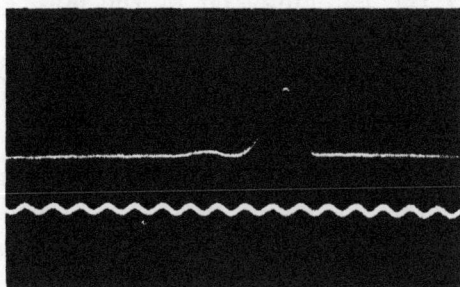

FIG. 40. OSCILLATION OF SECONDARY CIRCUIT OF TRANSFORMER

the maximum potential being in the lower curve 100,000 volts, in the upper curve 96,000 volts.

If the interrupter is used without a condenser, the curve shows the single oscillation of the secondary circuit alone, uninfluenced by the primary current. A curve obtained in this

way ($C_1 = 0$), from which the period and damping factor of
the secondary coil can be obtained, is shown in Fig. 40.

In order to obtain a comparison of the secondary voltage of
a transformer when supplied with alternating current with that

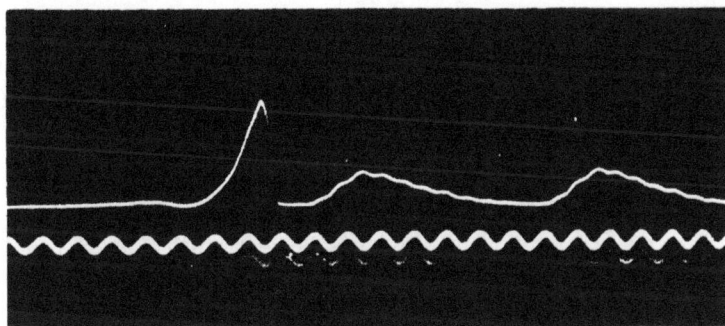

FIG. 41. TRANSFORMER WITH A.C. AND INTERRUPTER
Secondary voltage.

produced by interrupting the primary current the transformer
may be connected to an A.C. source through an interrupter.
Fig. 41 shows a curve obtained in this way, the A.C. supply
being at 205 volts, 4·6 amp. (R.M.S.), 50 ~, and the capacity
connected across the interrupter being 0·2 mfd. The curve

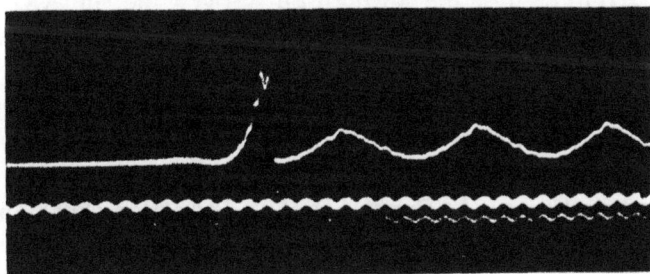

FIG. 42. TRANSFORMER WITH A.C. AND INTERRUPTER
Maximum A.C. voltage 76,500, peak potential at break 126,000 volts.

shows one complete alternation of the A.C. potential followed
by the potential at "break" which occurs almost when the
primary current is at its maximum value. In this case, the
maximum potential at break is 1·8 times the maximum A.C.
potential in the secondary circuit.

Another example is shown in Fig. 42, taken with a larger

primary capacity (1 mfd.). In this experiment, the maximum
A.C. secondary potential was 76,500 volts, the maximum after
"break" being 126,000 volts. In Fig. 43, the maximum potential after break is three times as great as the maximum A.C.
value.

When it is remembered that the closed-core transformer is,
as compared with an induction coil, very ineffective when used
with an interrupter, it is obvious that as a means of producing
high secondary potential the induction coil is far superior to the

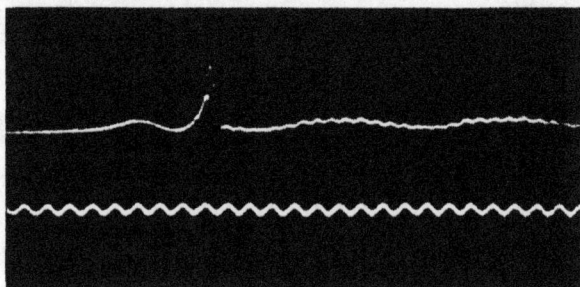

FIG. 43. TRANSFORMER WITH A.C. AND INTERRUPTER
Secondary voltage.

A.C. transformer. On the other hand, if it is desired to drive as
much current as possible through an X-ray tube the A.C. transformer is (for a given P.D. at the primary terminals) superior
to the coil. In the transformer, the reduction of the impedance
of the system due to the discharge calls up much more current
from the supply, an effect to which there is no parallel in the
action of induction coils.

Determination of the Constants of an Induction Coil. In order
to be able to calculate the secondary potential produced by an
induction coil at the interruption of a given primary current
from the constants of the circuits it is sufficient to know the
values of the six quantities $L_1 C_1$, $L_2 C_2$, k^2, L_{21}, R_1/L_1 and R_2/L_2.
From these quantities and the primary current may be cal-
culated the frequencies, initial amplitudes, damping factors,
and phases of the two oscillations of the coupled system, and
the secondary potential at any time after the interruption of
the primary current.

In determining the six constants, oscillation methods should be used as far as possible in order to ensure that the circumstances in which the measurements are made are similar to those existing in the working conditions of the coil. Some of the constants are determined by measuring frequencies of oscillation from photographed curves obtained with the electrostatic or the current oscillograph. If the frequencies are found by comparison with the tuning-fork curves, as explained on page 72, it is important to remember that a correction may be necessary owing to the fact that the oscillograph and fork spots sometimes do not cross the plate with the same velocity. If T is the period of the electrical oscillation to be determined, T' that of the tuning-fork, λ the wave-length of the oscillograph curve, λ' that of the corresponding portion of the tuning-fork curve, v and v' the velocities across the plate of the rays reflected from the oscillograph and the tuning-fork mirrors respectively, then $T = T'\lambda v'/\lambda' v$. The ratio of the velocities of the rays in any part of the plate can be found by photographing the two spots, and measuring the distance between them, when stationary in various positions on the plate.*

The coupling k^2 may be determined by finding the frequency of oscillation of the secondary circuit first with the primary coil open (the primary condenser being disconnected), then with the primary short-circuited. The ratio of the squares of these frequencies is $1 - k^2$. In this experiment the oscillations are photographed with the help of the electrostatic oscillograph connected to the secondary terminals, and are started by sparking to these terminals with an auxiliary coil.

The quantity $L_1 C_1$ is best found by removing the primary coil from the secondary and determining its period of oscillation with the current oscillograph which is connected in series with the coil, the battery, and the interrupter. Unless the damping of the oscillation is excessively great, the period is equal to $2\pi\sqrt{L_1 C_1}$. Another method for $L_1 C_1$ is to couple the primary coil very loosely with the secondary and to find the period with the electrostatic oscillograph connected to the secondary terminals. With a small correction for the reaction of the secondary current the expression for the period of the primary is $2\pi\sqrt{\{L_1 C_1 (1 + k^2 m)\}}$, where k^2 is the coupling in the

* *Phil. May.*, p. 243, (Aug., 1907).

circumstances of this experiment and m is the ratio L_2C_2/L_1C_1, both of these quantities being small.

With regard to L_2C_2 this quantity is best determined by reinserting the primary coil in the secondary, disconnecting the primary condenser, and connecting the secondary terminals to the electrostatic oscillograph. The oscillation being started by interrupting a suitable primary current, the period is given by the expression $2\pi\sqrt{L_2C_2}$, from which the value of L_2C_2 can be determined for the actual conditions as regards capacity and distribution of current which exist in the coil in working conditions. An example of a curve of secondary potential obtained by this method ($C_1 = 0$) is shown in Fig. 40, page 80.

The coefficient L_{21} of induction of the primary coil on the secondary, i.e the "mutual inductance" of the coils in the ordinary sense, may be determined by comparison with a standard mutual inductance by the ballistic galvanometer method. This experiment should be made with a number of different currents in the primary circuit in order to find the current at which L_{21} has its maximum value, the range of current over which L_{21} is practically constant, and its variation at other currents. If the primary self-inductance L_1, the coupling and the secondary self-inductance L_2 for uniformly distributed currents in the secondary coil are known, the value of L_{21} may also be found from the relation $L_{21} = k\sqrt{L_1L_2}$.

The ratios R_1/L_1, R_2/L_2, are determined from the logarithmic decrements of the oscillations used in determining L_1C_1 and L_2C_2. If γ is the ratio of the amplitudes at two successive peaks in one of these photographed curves, T the period of the oscillation, then R_1/L_1 (or R_2/L_2) is equal to $\frac{1}{2}T \log_e \gamma$.

Expressions for Secondary and Primary Potentials. The general expression for the secondary potential after "break" is of the form—

$$V_2 = A_1e^{-k_1t}\sin(2\pi n_1t - \delta_1) - A_2e^{-k_2t}\sin(2\pi n_2t - \delta_2),$$

where A_1, A_2 are the initial amplitudes, k_1, k_2 the damping factors, and $-\delta_1$, $-\delta_2$ the initial phases of the two component oscillations.

Unless the damping factors are excessively great the frequencies are given with sufficient accuracy by equation (13), page 6, and the initial amplitudes are those given in the expression (21), page 12, for the case when the resistances are negligible.

Employing a notation suggested by Drude[*], the expressions for the damping factors are—

$$k_1 = 4\pi^2 n_1{}^2 \left(\frac{\theta_1 + \theta_2}{2} - \beta \right) \qquad \qquad (46)$$

$$k_2 = 4\pi^2 n_2{}^2 \left(\frac{\theta_1 + \theta_2}{2} + \beta \right) \qquad \qquad (47)$$

in which $\theta_1 = \tfrac{1}{2} \dfrac{R_1}{L_1} \cdot L_1 C_1,$

$$\theta_2 = \tfrac{1}{2} \frac{R_2}{L_2} \cdot L_2 C_2,$$

and $\qquad \beta = - \dfrac{2\pi^2 n_1{}^2 n_2{}^2}{n_2{}^2 - n_1{}^2} \left(\theta_1 - \theta_2 \right) \left(L_1 C_1 - L_2 C_2 \right) \qquad . \qquad (48)$

The initial phases are given by—

$$\left. \begin{aligned} \tan \delta_1 &= 2\pi n_1 \left\{ 2\beta \frac{n_1{}^2 + n_2{}^2}{n_2{}^2 - n_1{}^2} - (\theta_1 - 2\theta_0 - \theta_2) \right\} \\[2mm] \tan \delta_2 &= \frac{n_2}{n_1} \tan \delta_1 \end{aligned} \right\} \qquad (49)$$

Where θ_0 is written for $\tfrac{1}{2} R_0 C_1$, R_0 being the ordinary ohmic resistance of the primary circuit, i.e. its resistance for steady currents.

The solution for the primary potential has been given by Dibbern[†] in the form—

$$V_1 = \frac{2\pi i_0 n_1 n_2{}^2}{C_1 (n_2{}^2 - n_1{}^2) \cos e} \cdot \left(L_1 C_1 - \frac{1}{4\pi^2 n_2{}^2} \right) e^{-k_1 t} \sin (2\pi n_1 t - e)$$

$$- \frac{2\pi i_0 n_1{}^2 n_2}{C_1 (n_2{}^2 - n_1{}^2) \cos e_1} \left(L_1 C_1 - \frac{1}{4\pi^2 n_1{}^2} \right) e^{-k_2 t} \sin (2\pi n_2 t - e_1) \quad (50)$$

where

$$\sin e = 2\pi n_1 R_0 C_1 + \frac{8\pi \beta n_1 n_2{}^2}{n_2{}^2 - n_1{}^2} + \frac{2\pi n_1 L_1 C_1 (\theta_1 - \theta_2 - 2\beta)}{L_1 C_1 - 1/4\pi^2 n_2{}^2},$$

$$\sin e_1 = 2\pi n_2 R_0 C_1 + \frac{8\pi \beta n_1{}^2 n_2}{n_2{}^2 - n_1{}^2} + \frac{2\pi n_2 L_1 C_1 (\theta_1 - \theta_2 + 2\beta)}{L_1 C_1 - 1/4\pi^2 n_1{}^2}.$$

[*] *Ann. d. Physik*, 13, p. 512 (1904).
[†] *Ann. d. Physik*, 40, p. 935 (1913).

Comparison of Calculated and Observed Wave Forms. Two examples are given here of the comparison of the calculated and observed wave forms of secondary potential,* the coil examined in both being that used in the experiments of Figs. 14 to 19, pages 47–53. In the first, the secondary terminals were connected only to the electrostatic oscillograph and a

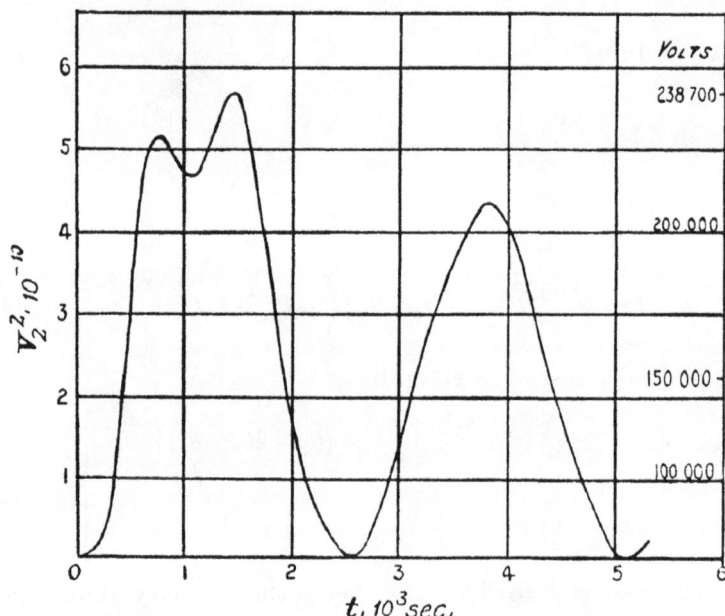

FIG. 44. CALCULATED SECONDARY VOLTAGE CURVE ($V_2{}^2$)

spark gap having spherical electrodes 2 cm. in diameter, and the primary capacity was 3 mfd.

The constants determined in these conditions were—

$$L_1 C_1 = 5\cdot822 \cdot 10^{-7} \text{ c.g.s.} \qquad k^2 = 0\cdot768$$

$$L_2 C_2 = 1\cdot114 \cdot 10^{-7} \text{ c.g.s.} \qquad R_1/L_1 = 149$$

$$L_{21} = 20\cdot4 \text{ henries} \qquad R_2/L_2 = 215.$$

The calculated frequencies were $n_1 = 194\cdot3$, $n_2 = 1{,}063$, and the calculated wave form of the secondary potential squared (for $i_0 = 10$ amp.) is shown in Fig. 44. The oscillograph curve

* Other examples of this comparison are given in *The Theory of the Induction Coil*, Chapter VII, and in *Phil. Mag.*, p. 28 (Jan., 1909); p. 706 (Nov., 1911); and p. 565 (April, 1914).

(taken at $i_0 = 3\cdot8$ amp.) is shown in Fig. 45. The agreement between the two curves shows that the frequencies, phases, and damping factors of the two component oscillations are well represented by the formulae given in the previous section. With regard to the maximum value of the secondary potential, the smallest primary current i_0 which was capable with this arrangement of giving at break a secondary spark 1 cm. long was found, when inserted in the theoretical expression for V_2, to give a maximum of 31,000 volts, in agreement with the

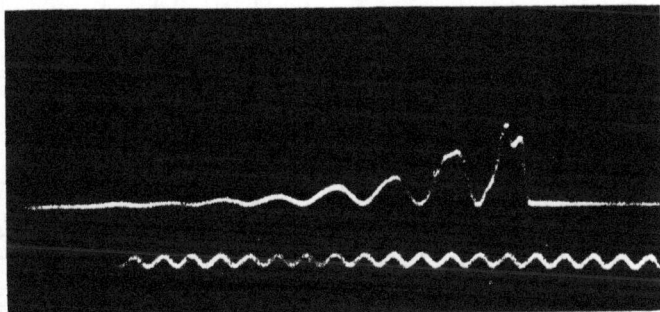

FIG. 45. OBSERVED SECONDARY VOLTAGE CURVE

generally accepted value for this spark length. The curve of sparking potentials given in Fig. 5, page 23, was determined by observations and calculations on this arrangement.

In the second example,* the coupling of the coil was reduced by the addition of series inductance (in the form of air-core coils) to the primary circuit, and the primary capacity was adjusted to its optimum value, $0\cdot2$ mfd. The constants for this case were—

$$L_1C_1 = 5\cdot11 \; . \; 10^{-9} \text{ c.g.s.} \qquad k^2 = 0\cdot560$$
$$L_2C_2 = 1\cdot142 = 10^{-7} \text{ c.g.s.} \qquad R_1/L_1 = 680$$
$$L_{21} = 20\cdot4 \text{ henries} \qquad R_2/L_2 = 825.$$

The frequencies were $n_1 = 413\cdot9$, $n_2 = 1,208$, and the system was nearly in the first of the adjustments of Table I, page 32, the frequency ratio being $2\cdot919$, and the value of $u(= L_1C_1/L_2C_2)$ being $0\cdot447$.

In Fig. 46 (A) are shown the two component oscillations in the wave of secondary potential separately (calculated for

* Phil. Mag., p. 1 (Jan., 1915).

$i_0 = 10$ amp.), and Fig. 46 (B) is the result of their superposition. It will be seen that two positive maxima occur almost simultaneously at about $t = 0\cdot0006$ sec., giving rise to the maximum secondary potential (596,900 volts) at this instant. Fig. 47 is the calculated $V_2{}^2$ curve, and Fig. 48 the corresponding

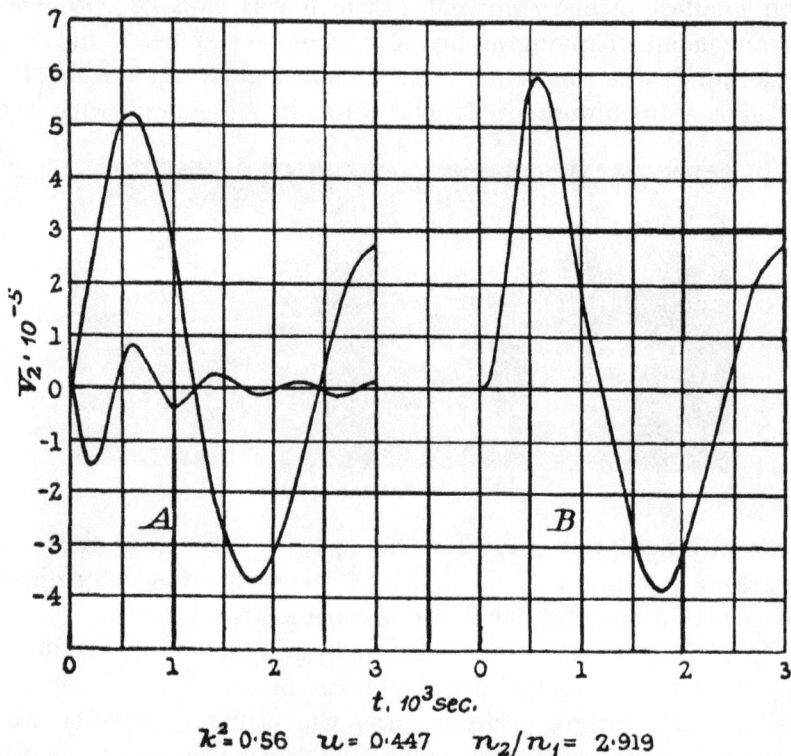

$$k^2 = 0\cdot56 \quad u = 0\cdot447 \quad n_2/n_1 = 2\cdot919$$

Fig. 46

A = Component oscillations in secondary.
B = Resultant oscillation in secondary.

oscillogram taken at $i_0 = 2$ amp. In both of these curves, the waves have the peaked summits and flattened zeroes which characterize the 3/1 frequency ratio. The curves also agree well in period and rate of decay. The calculated maximum at $i_0 = 2$ amp. is 116,000 volts, the observed spark length at this current being 18·2 cm.

The *primary* potential was also calculated for this case by Dibbern's formula (see page 85), the result being shown in Fig. 49. In this diagram A represents the two primary potential oscillations separately, B the result of their superposition. The initial amplitudes of the two primary potential components are nearly equal, as required by the condition $u = 1 - k^2$, which

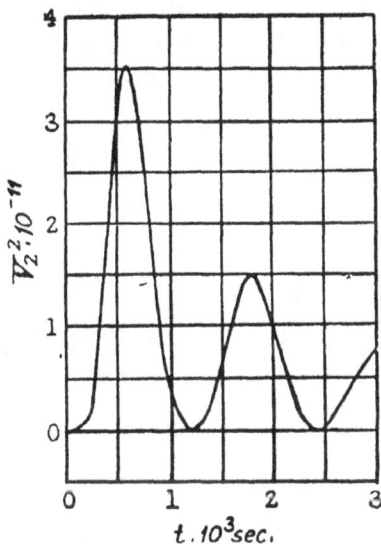

Fig. 47. CALCULATED SECONDARY VOLTAGE CURVE ($V_2{}^2$)

is here approximately fulfilled. It will be seen from Fig. 49 (B) that the potential of the primary condenser reaches a maximum of 6,800 volts at about 0·00025 sec., and a minimum of 2,250

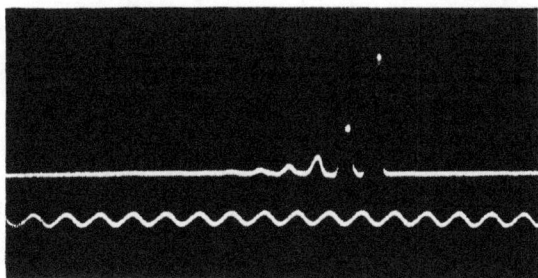

Fig. 48. OBSERVED SECONDARY VOLTAGE CURVE

volts at about 0·00065 sec. after the interruption. Thus, at the moment at which the secondary potential reaches its greatest value the primary condenser, instead of being uncharged, as would be the case if the damping coefficients of the two oscillations were zero or equal (see Fig. 7, page 34), is still charged to about 2,250 volts. The effective resistances therefore act in two ways in reducing the conversion efficiency of the arrangement. First, they give rise to dissipation of energy and consequent decay of both oscillations. Second, owing to the *difference* between the damping factors of the two oscillations,

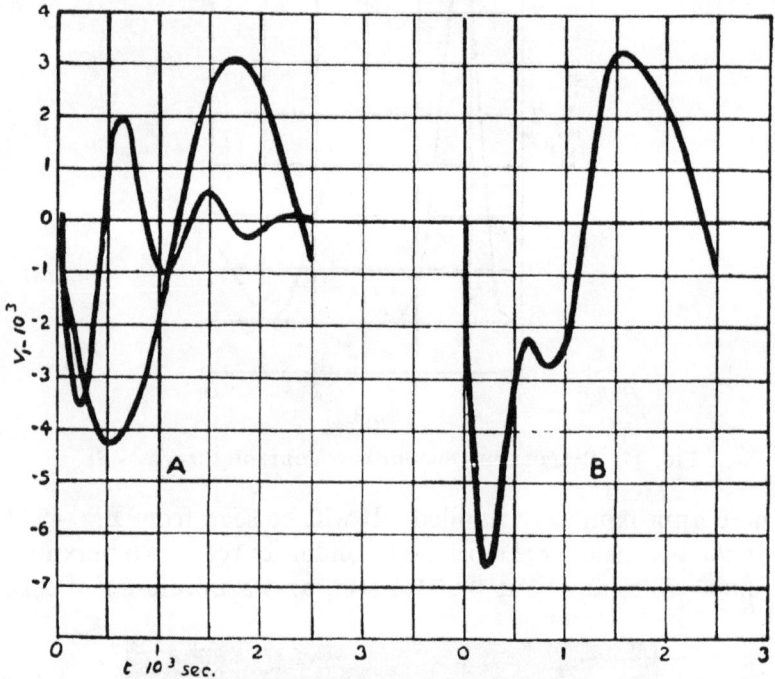

FIG. 49

A - Component oscillations in primary.
B = Resultant oscillation in primary.

there is some energy stored in the primary condenser at the moment when the secondary potential is at its maximum.

The conversion efficiency may be calculated as follows: The maximum secondary potential ($i_0 = 10$ amp.) is 596,900 volts. If the resistances were negligible, and if the system were exactly in the first unit-efficiency adjustment, the potential would be, by (32), page 31, i.e. $V_{2m} = L_{21}i_0/k\sqrt{L_2 C_2}$— 798,700 volts. Consequently, the conversion efficiency is $\left(\dfrac{5969}{7987}\right)^2$ $= 0.559$. Since $L_1 = 0.255$ henry, the initial magnetic energy

$\frac{1}{2}L_1 i_0^2$ is 12·75 joules. The maximum electrostatic energy in the secondary circuit is therefore $0·559 \times 12·75$, or about 7·1 joules. The energy at the same moment in the primary condenser (capacity 0·2 mfd.) is $\frac{1}{2}$ 0·2 . 2250^2 . 10 ergs, or about 0·5 joule. Consequently, of the original 12·75 joules, rather over 5 are dissipated, half a joule is stored in the primary condenser, and the remaining 7 joules represent the electrostatic energy of the secondary charge at the moment of maximum potential. Taking the secondary capacity as 0·000053 mfd. (see page 93) the charge then accumulated on the secondary is $5·3 \times 5·969 \times 10^{-7}$ c.g.s. units, or $31·6$. 10^{-6} coulombs.

Determination of the Capacity of the Secondary Coil. When the secondary terminals are connected with bodies of very small capacity, C_2 consists mainly of the capacity distributed over the secondary coil. It is defined, as previously stated, so that $C_2 V_2$ is the secondary charge, V_2 being the potential difference of the terminals, and so that $2\pi\sqrt{L_2 C_2}$ is (neglecting resistance) the period of the secondary circuit when oscillating alone, e.g. when the primary circuit is broken and disconnected from the condenser. During the oscillations the secondary current is not uniformly distributed over the length of the wire, and the self-inductance L_2 is considerably less than the value which this quantity would have if the current were the same in all windings, as would be the case if the secondary terminals were connected with a condenser of large capacity. While the product $L_2 C_2$ may be readily determined from the period, the determination of L_2 and C_2 separately presents much greater difficulties. As an illustration of an approximate method, the following account is given of a determination of these quantities for an 18 in. coil, the secondary terminals of which were at the time connected to a pair of spark electrodes and to the electrostatic oscillograph—bodies of much smaller capacity than that of the secondary coil itself.

The method requires the determination of the quantities L_1, L_{21}, k^2, $L_2 C_2$, and the ratio L_{21}/L_{12}, from which C_2 can be calculated by the equation

$$C_2 = \frac{k^2 L_1 . L_2 C_2}{L_{21}^2} . \frac{L_{21}}{L_{12}} \qquad \qquad \qquad . \quad (51)$$

The first four of these quantities were determined by methods described earlier in the present Chapter. With regard to the ratio L_{21}/L_{12}, this differs from unity because the current during the oscillations is not uniform along the length of the secondary coil, but is greatest at the centre and nearly zero at the ends. If we assume as an approximation that the current in a secondary winding at distance z from the middle is proportional to $\cos \dfrac{\pi z}{h}$, where h is the length of the secondary coil, and if all the windings of the secondary had equal inductive effects on the primary when reckoned per unit current, it is easily seen that L_{21}/L_{12} would be equal to $\pi/2$. In the actual case, however, the inductive effect of the secondary windings (per unit current) diminishes from the centre towards each end. This was tested by ballistic galvanometer experiments in which the mutual inductance of the primary and a single turn of wire, wound on the secondary (or primary) in various positions, was compared with its value for the central position. From the results of these measurements, it was found that this mutual inductance could be represented approximately by the expression $a - bz^2 - cz^4$. Consequently, L_{12} is proportional to

$$\int_{-h/2}^{+h/2} (a - bz^2 - cz^4) \cos \frac{\pi z}{h}\, dz,$$

while L_{21} is proportional to

$$\int_{-h/2}^{+h/2} (a - bz^2 - cz^4)dz,$$

since the current in the primary coil is uniformly distributed. The value of L_{21}/L_{12} is thereby reduced from $\pi/2$, and becomes in the present case $0 \cdot 95\pi/2$.

Another correction is necessary if, as in the present experiments, the secondary terminals are connected with a capacity which is not negligible in comparison with that of the coil. In this case the secondary current is not quite zero at the ends of the coil, but should be represented as proportional to $\cos \dfrac{\pi z}{h'}$, where h' is greater than h. The value of h' may be estimated if we know the ratio of the external capacity C_e to the total capacity C_2. If C_e is small in comparison with C_2 the approximate value of the latter (obtained from equation (51) by neglecting the present correction) may be used here.

If the current in the secondary windings varies as $\cos \frac{\pi z}{h'}$, the

charge per unit length will be proportional to $\sin \frac{\pi z}{h'}$. Hence the ratio of C_e to C_2 is equal to the ratio of

$$\int_{h/2}^{h'/2} \sin \frac{\pi z}{h'} . \, dz \text{ to } \int_0^{h'/2} \sin \frac{\pi z}{h'} dz,$$

i.e.
$$\frac{C_e}{C_2} = \cos \frac{\pi h}{2h'}.$$

This determines h'/h, and we then have

$$\frac{L_{12}}{L_{21}} = \frac{1}{h} \int_{h/2}^{+h/2} \cos \frac{\pi z}{h'} dz$$

$$= \frac{2}{\pi} . \frac{h'}{h} \sin \frac{\pi h}{2h'}.$$

In the present experiments, C_e is the capacity of the oscillograph and the spark-gap terminals, and this is about one-sixth of the total secondary capacity C_2, the value of which is already known approximately. Hence, $\cos \frac{\pi h}{2h'} = \frac{1}{6}$, and $\frac{h'}{h} = 1.12$,

from which $\frac{L_{12}}{L_{21}} = 1.10 . \frac{2}{\pi}$. The effect of this correction is therefore further to reduce L_{21}/L_{12} by about 10 per cent.

Taking both corrections into account, we have approximately

$$L_{21}/L_{12} = 0.85\pi/2 = 1.335,$$

from which by (51), putting in the values $k^2 = 0.768$, $L_1 = 0.194$ henry, $L_{21} = 20.4$ henries, $L_2 C_2 = 1.114 . 10^{-7}$ c.g.s.—all determined for approximately the same mean magnetization of the core—we find

$$C_2 = 0.000053 \text{ mfd.},$$

and with the above value of the product $L_2 C_2$,

$$L_2 = 2,150 \text{ henries.}$$

When the secondary terminals were connected with a condenser of 0·001 mfd. capacity—in which case the oscillating current can be regarded as practically uniformly distributed—the value found for the self-inductance was 2,540 henries. It appears therefore, that the self-inductance of the secondary coil is reduced, by the change in the distribution of the current, in a smaller ratio than is the inductance of the secondary on the primary—a result which agrees with the observation that the coefficient of coupling is also reduced by the change from the uniform to the non-uniform distribution of current.

CHAPTER IV

THE PRIMARY CIRCUIT. THE SECONDARY POTENTIAL
AT "MAKE"

In Fig. 50 is shown a typical oscillogram of the primary current at "break." The curve was taken with the current oscillograph connected in the primary circuit, the break being effected by the "slow" interrupter operated by the rotating mirror. The deflexion of the curve from the zero line is proportional to the primary current, the steady current i_0 before break

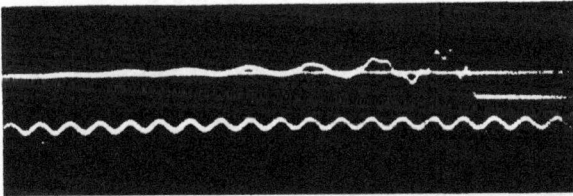

FIG. 50. OSCILLOGRAM OF PRIMARY CURRENT AT BREAK

being represented by a deflexion downwards. The curve shows the manner in which the primary current falls to zero after break. It consists of two oscillatory components, the frequency ratio in this case being 4. In the experiment of Fig. 50 there was no condenser in the secondary circuit, and no discharge between the secondary terminals.

Fig. 51 shows two oscillograms of the primary current obtained with a motor interrupter substituted for the slow contact breaker. In this experiment the connections of the oscillograph were reversed so that the deflexion due to the battery current is upwards. In the curves of Fig. 51, the lines sloping upwards from right to left, having a superposed oscillation in their early portions, represent the growth of the primary current after make. The oscillations immediately after break are similar to those of Fig. 50. In the upper curve of Fig. 51, the primary capacity was larger, the period of the main oscillation after break consequently longer and the frequency ratio greater, than in the lower curve. The frequency of the oscillation after "make" is, of course, not affected by the primary capacity, which is shortcircuited at make. In the interrupter used in the experiments of Fig. 51, "make" occurs very soon after break, usually after about one complete oscillation of the slower component. The interrupter was, however, not working very regularly, "make" occurring sometimes sooner after "break," e.g. in the first wave on the right in the lower curve. When the photographs which are reproduced in Fig. 51 were

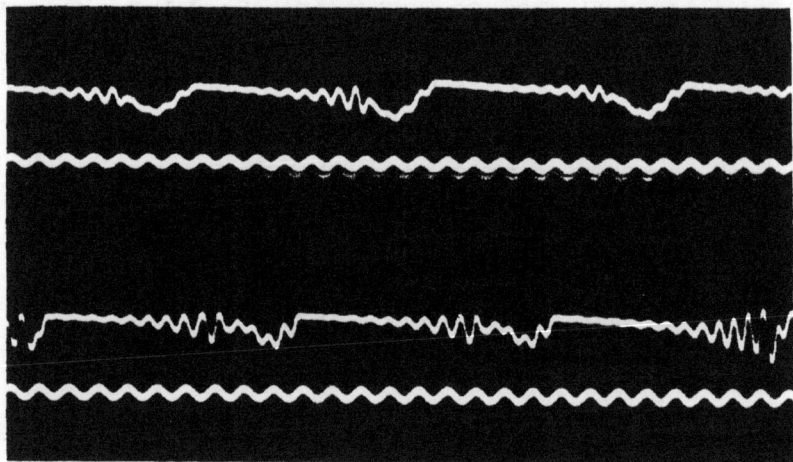

FIG. 51. OSCILLOGRAMS OF PRIMARY CURRENT

taken, short sparks were passing between the secondary terminals, but this form of discharge does not appear to affect appreciably the wave form of the primary current.

In the action of induction coils the manner of variation of current in the primary circuit after break is only indirectly of importance, the two most important quantities in this circuit being the maximum potential of the primary condenser after break, and the rate of growth of the primary current after make. Upon the first depends the strength of the dielectric required between the plates of the condenser, and the second determines, when a rapid interrupter is used, the current i_0 at break, which is directly proportional to the maximum secondary potential. These two quantities we will, therefore, now proceed to consider.

The Maximum Primary Potential. It has sometimes been assumed that the maximum secondary and primary potentials developed at the interruption of the primary current in an induction coil are related according to the usual transformer law, viz. that their ratio is equal to the ratio of the numbers of turns in the secondary and primary coils. Calculation shows, however, that this relation holds only in certain exceptional cases. In general, the ratio of the potentials depends also upon the coupling and the capacities of the system.

For the purpose of this calculation, the resistances of the circuits may be neglected, so that the frequencies are given by equation (13), page 6, the potentials by equations (21) and (22), page 12.

In the first place, it is easily seen that the transformer law is applicable when there is no magnetic leakage ($k^2 = 1$) whatever be the capacities associated with the primary and secondary coils. In this case the frequencies are

$$n_1 = \frac{1}{2\pi\sqrt{(L_1C_1 + L_2C_2)}}, \quad n_2 = \infty .$$

In each circuit there is only one oscillation, the amplitude of the infinitely rapid component being equal to zero. The expressions for the potentials become

$$V_1 = -\frac{L_1 i_0}{\sqrt{(L_1C_1 + L_2C_2)}} \sin 2\pi n_1 t,$$

$$V_2 = \frac{L_{21} i_0}{\sqrt{(L_1C_1 + L_2C_2)}} \sin 2\pi n_1 t.$$

The ratio of the maximum secondary to the maximum primary potential is thus L_{21}/L_1, which, since there is no magnetic leakage, is equal to the ratio of the numbers of turns in the secondary and primary coils. This result would hold even if the resistances were taken into account, since the primary and secondary oscillations have equal damping factors and their maxima occur at the same value of t.

The transformer law therefore holds accurately when $k^2 = 1$, but in practice no induction coil has so high a coefficient of coupling. In coils of the usual form with straight cores, k^2

probably never exceeds, though it may nearly reach, the value 0·9. It therefore becomes necessary to consider the ratio of the potentials in cases in which there is appreciable magnetic leakage so that the frequency-ratio is finite and both oscillations are called into play.

Still neglecting the resistances, we may calculate the maximum secondary potential in any such case from the expression (42), page 37,

$$V_{2m} = \frac{L_{21}i_0}{\sqrt{L_2 C_2}} U \sin \phi.$$

There is, however, no such simple equation for the maximum *primary* potential, the value of which must be arrived at by numerical calculation from the expression for V_1. For this purpose the expression (22), page 12, may conveniently be written

$$V_1 = \frac{L_1 i_0}{2\sqrt{L_2 C_2}} (a_1 \sin 2\pi n_1 t - a_2 \sin 2\pi n_2 t),$$

where

$$a_1 = \frac{U}{1 + \frac{n_1}{n_2}} \cdot \frac{1 - u - \sqrt{\{(1-u)^2 + 4k^2 u\}}}{u},$$

$$a_2 = \frac{U}{1 + \frac{n_2}{n_1}} \cdot \frac{1 - u + \sqrt{\{(1-u)^2 + 4k^2 u\}}}{u},$$

and u, as before, represents the ratio $L_1 C_1 / L_2 C_2$.

The function $a_1 \sin 2\pi n_1 t - a_2 \sin 2\pi n_2 t$ being denoted by ψ the problem is to determine for any values of k^2 and u the maximum value ψ_m of this function in the first half-period of the slow oscillation. As in the case of the secondary potential, we need not consider what happens in subsequent phases, since the potential is then so much reduced in practice by the damping as to be always less than the greatest in the first half-period.

By numerical calculation of the frequency ratios and of U, a_1, a_2, values of ψ_m were determined for a number of values of u and for the two couplings $k^2 = 0·9$ and $k^2 = 0·571$. The results are given in the second and sixth columns of Table III, for the values of u shown in the first and fifth columns. The third and seventh columns contain the values of $U \sin \varphi$, the fourth and eighth those of the ratio $U \sin \varphi / \psi_m$

TABLE III

$k^2 = 0.9$				$k^2 = 0.571$			
u	ψ_m	$U \sin \varphi$	$\dfrac{U \sin \varphi}{\psi_m}$	u	ψ_m	$U \sin \varphi$	$\dfrac{U \sin \varphi}{\psi_m}$
0·05	4·391	1·047	0·2384	0·1	4·988	1·070	0·2145
0·1	3·443	1·054	0·3061	0·2	3·552	1·229	0·3460
0·2	2·792	1·016	0·3639	0·3	2·739	1·300	0·4745
0·4	2·099	0·9995	0·4761	0·429	2·035	1·323	0·6501
0·6	1·868	0·9408	0·5036	0·6	1·734	1·302	0·7508
0·8	1·700	0·8865	0·5214	0·8	1·518	1·251	0·8241
1·0	1·570	0·8407	0·5356	1·0	1·371	1·192	0·8696
5·0	0·8348	0·4542	0·5441	5·0	0·8352	0·5495	0·6579
10·0	0·6006	0·3334	0·5551	10·0	0·6200	0·3384	0·5458

The ratio of the maximum secondary to the maximum primary potential is

$$\frac{V_{2m}}{V_{1m}} = \frac{2L_{21}}{L_1} \cdot \frac{U \sin \varphi}{\psi_m} \qquad \qquad . \qquad . \qquad . \qquad . \qquad . \quad (52)$$

When u is varied by changing only the primary capacity, the quantity $2L_{21}/L_1$ is constant, and $U \sin \varphi / \psi_m$ is then *proportional* to the ratio of the maximum potentials. Table III shows that even when the coupling is as high as 0·9 ($k = 0.949$) these potentials are by no means in a constant ratio. When the primary capacity is very small, V_{2m}/V_{1m} is much smaller than L_{21}/L_1, which may, with the usual approximation, be taken as equal to the ratio of the numbers of secondary and primary turns. As the primary capacity is increased V_{2m}/V_{1m} increases somewhat rapidly and becomes equal to the ratio of the turns when u is about 0·6, i.e. when $U \sin \varphi / \psi_m$ is 0·5. The primary capacity being further increased the potential ratio varies more slowly, only reach'ng the value $1\cdot110 L_{21}/L_1$ when $u = 10$.

At the looser coupling $k^2 = 0.571$ the variation of the

potential-ratio as the primary capacity is increased is more pronounced. It rises to $1\cdot739 L_{21}/L_1$ at $u = 1$ ($L_1 C_1 = L_2 C_2$), subsequently falling to $1\cdot0916 L_{21}/L_1$ at $u = 10$. At this degree of coupling, however, L_{21}/L_1 cannot be taken as approximately equal to the ratio of the secondary and primary turns.

It appears that if the primary capacity is very much increased the potential-ratio again becomes equal to L_{21}/L_1; in other words, when u is indefinitely increased $U \sin \varphi/\psi_m$ becomes equal to $0\cdot5$ for all values of k^2. For in this case the frequencies are $n_1 = 1/2\pi\sqrt{L_1 C_1}$, which is very small, and $n_2 = 1/2\pi\sqrt{\{L_2 C_2(1 - k^2)\}}$, which is equal to the frequency of the secondary circuit with the primary closed. The frequency-ratio n_2/n_1 being very great, the amplitude of the rapid oscillation is in each circuit negligible, so that the system again oscillates with practically only one frequency, this being n_1. The amplitudes are—in the secondary $L_{21} i_0 U/\sqrt{L_2 C_2}$, in the primary $L_1 i_0 U/\sqrt{L_2 C_2}$, the potential-ratio V_{2m}/V_{1m} being therefore equal to L_{21}/L_1. This case is, however, of no practical importance, since, owing to the small value of U, no high secondary potential can be developed by an induction coil provided with a very large primary capacity.

The present theory therefore indicates that, while the transformer law holds in the two extreme cases $k^2 = 1$ and C_1 very great, it is not generally applicable to an induction coil in actual working conditions. It appears, however, from the values given in Table III that the potential-ratio is equal to L_{21}/L_1 at a certain value of u which depends upon the coupling, e.g. $0\cdot6$ if $k^2 = 0\cdot9$, and rather over $0\cdot3$ if $k^2 = 0\cdot571$.

In Table IV are shown the values of ψ_m for a few other adjustments involving different degrees of coupling, the value of u in each case being that corresponding to the "optimum" primary capacity.

Within the range of this table—probably the whole of the practical range of coupling found in induction coils—if the primary capacity is adjusted to its optimum value, the numerical value of ψ_m appears to lie between 2 and 4, the exact value depending on the coupling. For a given degree of coupling ψ_m diminishes as the ratio $u(= L_1 C_1/L_2 C_2)$ is increased (see Table III), but much less rapidly than according to the law of inverse proportion; it is more nearly inversely proportional to the square root of u.

TABLE IV

VALUES OF ψ_m IN "OPTIMUM" ADJUSTMENTS

k^2.	u	ψ_m
0·9	0·1	3·443
0·835	0·165	3·101
0·768	0·11	3·973
0·71	0·44	2·297
0·64	0·45	2·131
0·571	0·429	2·035

According to the above calculation, the expression for the maximum potential difference of the plates of the primary condenser, arrived at on the supposition that the oscillations are undamped and that no discharge takes place either at the interrupter or between the secondary terminals, is

$$V_{1m} = \frac{L_1 i_0}{2\sqrt{L_2 C_2}} \cdot \psi_m \qquad . \qquad . \qquad . \qquad . \qquad (53)$$

As a numerical example, we may consider the case illustrated in Figs. 46 to 49, pages 88–90, in which L_1 is 0·255 henry, $L_2 C_2$ is $1·142 \cdot 10^{-7}$ c.g.s., and the value of ψ_m (from Table III or IV) is 2·035. With $i_0 = 10$ amp., equation (53) gives for V_{1m} the value 7,680 volts. When the effect of the resistances is taken into account, however, the value of V_{1m} is (see page 89) 6,800 volts. The difference represents the effect of the damping during the short interval of time 0·00025 sec. which the primary potential requires to attain its maximum value.

By equation (52) the ratio of the maximum secondary to the maximum primary potential would be in this case, if the resistances were negligible,

$$\frac{V_{2m}}{V_{1m}} = \frac{2L_{21}}{L_1} \cdot \frac{U \sin \varphi}{\psi_m}$$

$$= 160 \times 0·65 \quad \text{(See Table III.)}$$

$$= 104.$$

With the resistances taken into account, the ratio is (see page 90)—

$$\frac{V_{2m}}{V_{1m}} = \frac{596,900}{6800}$$

$$= 87\cdot8.$$

Thus, the damping has a greater effect in lowering the maximum secondary than in reducing the maximum primary potential, which is to be expected since the maximum potential occurs later, and the damping forces are in action for a longer time before the maximum is reached, in the secondary circuit than in the primary.

The primary potential at break is, of course, greatest for a given capacity and current when the secondary coil is altogether removed, as indicated by the explosive nature of the interrupter spark in these conditions even at moderate currents. In this case, the principle of energy gives (resistances being neglected) $C_1 V_{1m}^2 = L_1 i_0^2$, so that $V_{1m} = i_0\sqrt{L_1/C_1}$. In the numerical example just considered (with the secondary coil removed) this gives 11,290 volts as the maximum potential of the condenser, but with allowance for the resistance the maximum is (by equation (7), page 4) 10,000 volts. Thus, the presence of the secondary coil reduces the maximum primary potential from 10,000 to 6,800 volts; in other words, the secondary coil withdraws more than one-half of the original energy from the primary circuit before the maximum potential in this circuit is reached, thus greatly facilitating the smooth working of the interrupter.

The method of producing high potentials with a single coil does not appear to have been much tried. If a suitable interrupter could be devised—it would require to be in a high vacuum, e.g. an X-ray tube—single-flash discharges of very high potential could be produced in this way. A coil, for example, having self-inductance 5 henries and capacity 0·00005 mfd., would develop (allowing a reduction of 15 per cent for damping) a potential of 270,000 volts at the interruption of a current of 1 amp., a performance much superior to any that can be expected in an induction coil of the usual construction.

Growth of the Primary Current After "Make." The value of the primary current at any time t after "make" is given approximately by the expression—

$$i_1 = \frac{E}{R_0}\left(1 - e^{-\delta t}\right) \qquad . \qquad . \qquad . \qquad . \qquad . \qquad . \qquad (54)$$

where E is the battery E.M.F., R_0 is the steady-current resistance of the primary circuit, and the coefficient δ may be taken as equal to R_0/L_1. Strictly speaking, there is also, owing to the reaction of the secondary coil, an oscillatory component in the primary current at make (see Fig. 51, page 96), and the coefficient of t is not exactly equal to R_0/L_1, but the oscillation usually dies out before break and the difference between δ and R_0/L_1 is not very considerable.

If T is the interval between make and break, i.e. the "time of contact," the current at break is therefore

$$i_0 = \frac{E}{R_0}\left(1 - e^{-R_0 T/L_1}\right) \qquad . \qquad . \qquad . \qquad . \qquad . \qquad (55)$$

If T_1 is the time interval between two successive breaks, the mean current, as indicated by an amperemeter in the primary circuit is—

$$i_m = \frac{1}{T_1}\int_0^T i_1 dt$$

$$= \frac{E}{R_0 T_1}\left[T - \frac{L_1}{R_0}\left(1 - e^{-R_0 T/L_1}\right)\right] \qquad . \qquad . \qquad . \qquad (56)$$

The work done in establishing the current i_0 is—

$$W = E i_m T_1$$

$$= \frac{E^2}{R_0}\left[T - \frac{L_1}{R_0}\left(1 - e^{-R_0 T/L_1}\right)\right]$$

The ratio of the electrokinetic energy $\frac{1}{2}L_1 i_0^2$ to the work done in establishing it—the "efficiency of make" as it has been called—is, therefore, equal to—

$$\frac{1}{2}\frac{L_1}{R_0} \cdot \frac{(1 - e^{-R_0 T/L_1})^2}{T - \frac{L_1}{R_0}(1 - e^{-R_0 T/L_1})}$$

or $\frac{1}{2}\dfrac{(1 - e^{-\nu})^2}{y - (1 - e^{-\nu})}$.   .   .   .   .   .   (57)

if y is written for $R_0 T/L_1$.

The efficiency of make is increased when y is diminished, that is, when the ratio L_1/R_0 is increased. If ample E.M.F. is available and the coil is operated through a series regulating resistance, the self-inductance L_1 can be increased, by the inclusion of inductance coils in the primary circuit, without increasing R_0. In this case, the addition of such coils will improve the efficiency of make. On the other hand, it will, by (55), diminish the current at break. It has been explained in Chapter II, however, that the action of a coil at break, in converting primary magnetic into secondary electrostatic energy, can be improved by the addition of series inductance to the primary circuit. On the whole, taking these various effects into account, it is found that, in such cases, the additional primary inductance, if suitably chosen, improves the performance of a coil.

As a numerical example of these calculations, we may consider the experiment illustrated by Fig. 31, page 75. In this case, E was 94 volts, R_0 was 18 ohms, L_1 0·255 henry, T_1 0·030 sec., and T was 0·00814 sec. By (55) the primary current i_0 at break was 2·28 amp., and by (56) the mean primary current was 0·339 amp. The current indicated by an ampere-meter in the primary circuit was 0·3 amp. The value of y in this experiment was 0·575, and, therefore, that of $e^{-\nu}$ was 0·563. By (57) the efficiency of make was, therefore, 0·692. The efficiency of conversion in this case was 0·559 (see page 90). If the spark gap was set so that sparks just passed (they were 18·4 cm. long) the resulting oscillogram of the secondary potential was that shown in Fig. 66, p. 127. From a comparison of this curve with that of Fig. 48, page 89, we find that about four-fifths of the energy of the system was dissipated in the spark, the remaining one-fifth representing the energy of the oscillations in the system subsequent to the appearance of the spark. Consequently, of the work done by the battery in establishing the magnetic energy $\frac{1}{2}L_1 i_0^2$ about 0·692 × 0·559 × 0·8, i.e. about 31 per cent, was dissipated in the spark between the secondary terminals.

The Secondary Potential at "Make."* The manner of variation of the primary current after contact is made at the interrupter is closely connected with that of the secondary potential
at make, and both of these quantities may be calculated (the
former with greater accuracy than was done in the previous
section) from the following equations. The chief interest of
the calculation of the secondary potential at make lies in the
means which the result may suggest for the diminution of this
quantity. In all applications of induction coils it is desired
to make the secondary discharge unidirectional, and it is,
therefore, necessary to suppress as far as possible the discharge
at make which is in the opposite direction to that at break.
For this purpose mechanical or valve rectifiers are sometimes
used, but all these absorb some energy and tend to detract
from the performance of the coil at break. It is, therefore,
desirable to consider how the circuits of an induction coil may
be adjusted so as to lower the potential at make, and in doing
so, due regard should be had to the necessity of maintaining
undiminished the potential at break.

In the following calculation, we shall regard the secondary
terminals as insulated, and shall suppose that immediately
before "make" there is no current flowing in either circuit.
The latter condition is usually satisfied, the oscillations at
break completely dying out before make (see, for example,
Fig. 31, page 75), though there are exceptional cases (see Figs.
37, page 78, and 51, page 96) in which "make" follows so
soon after "break" that there may be current flowing, one
way or the other, at the moment of "make."

The primary and secondary currents being i_1, i_2, and with
the usual notation for the other quantities, the equations for
the circuits are—

$$L_1 \frac{di_1}{dt} + L_{12} \frac{di_2}{dt} + R_1 i_1 = E, \qquad . \qquad . \qquad . \qquad . \qquad (58)$$

$$L_2 \frac{di_2}{dt} + L_{21} \frac{di_1}{dt} + R_2 i_2 + V_2 = 0, \qquad . \qquad . \qquad . \qquad (59)$$

$$i_2 = C_2 \frac{dV_2}{dt} \qquad . \qquad . \qquad . \qquad . \qquad . \qquad . \qquad (60)$$

---

* See also *The Theory of the Induction Coil*, Chapter X, and *Phil. Mag.*,
33, p. 322 (1917).

The primary capacity, which becomes short circuited at "make," has, of course, no influence on the current after this moment, and does not appear in the equations.

Denoting $i_1 - E/R_1$ by x, and making use of (60), we find equations (58) and (59) to become

$$L_1\frac{dx}{dt} + L_{12}C_2\frac{d^2V_2}{dt^2} + R_1x = 0, \qquad . \qquad . \qquad . \qquad (61)$$

$$L_2C_2\frac{d^2V_2}{dt} + L_{21}\frac{dx}{dt} + R_2C_2\frac{dV_2}{dt} + V_2 = 0. \qquad . \qquad . \qquad (62)$$

The solution of equations (61), (62) is of the form $x = Ae^{zt}$, $V_2 = Be^{zt}$, where z is a root of the cubic equation

$$(L_1L_2 - L_{12}L_{21})z^3 + (L_1R_2 + L_2R_1)z^2$$
$$+ \left(\frac{L_1}{C_2} + R_1R_2\right)z + \frac{R_1}{C_2} = 0. \qquad . \qquad . \qquad (63)$$

In all actual cases this equation will have two roots of the form

$$\left.\begin{array}{l}z_1 = -k_1 + 2\pi ni \\ z_2 = -k_1 - 2\pi ni\end{array}\right\} \qquad . \qquad . \qquad . \qquad (64)$$

where $i = \sqrt{-1}$, the third root being real, say,

$$z_3 = -\delta. \qquad . \qquad . \qquad . \qquad . \qquad . \qquad (65)$$

The complete solution of (61), (62) is thus

$$\left.\begin{array}{l}x = A_1e^{z_1t} + A_2e^{z_2t} + A_3e^{z_3t} \\ V_2 = B_1e^{z_1t} + B_2e^{z_2t} + B_3e^{z_3t}\end{array}\right\} \qquad . \qquad . \qquad . \qquad (66)$$

where the A's and B's are to be determined from the initial conditions. These are $i_1 = 0$, $V_2 = 0$, $i_2 = 0$, that is

$$x = -E/R_1, \ V_2 = 0, \ \frac{dV_2}{dt} = 0, \text{ when } t = 0. \qquad . \qquad . \qquad (67)$$

On substituting the expressions (66) for x and V_2 in (61) and (62) and eliminating A_1, A_2, A_3, we find, with the help of (67) the three equations for the coefficients B,

$$\left.\begin{array}{l}B_1 + B_2 + B_3 = 0 \qquad . \qquad . \qquad . \qquad . \qquad . \\ B_1z_1 + B_2z_2 + B_3z_3 = 0 . \qquad . \qquad . \qquad . \qquad . \\ B_1z_1^2 + B_2z_2^2 + B_3z_3^2 = -EL_{21}/C_2L_1L_2(1-k^2) .\end{array}\right\} \qquad (68)$$

Solving these for B_1, B_2, B_3, and retaining the real parts of the three terms of (66) we find finally, as the complete solution for V_2,

$$V_2 = \frac{EL_{21}}{L_1}\left\{ e^{-k_1 t} \cos (2\pi n t - \theta) - e^{-\delta t} \right\} \qquad . \qquad . \qquad . \qquad (69)$$

where $\tan \theta = \dfrac{k_1 - \delta}{2\pi n}$.

The wave of potential in the secondary circuit at make thus consists of an oscillation, having frequency n and damping factor k_1, superposed upon an exponentially decaying part, the initial values of the two parts being $\pm EL_{21}/L_1$. The greatest numerical value of the expression (69) for V_2 occurs at a time t_1 somewhat less than $1/2n$, and will in all cases be less than $2EL_{21}/L_1$, the value of the maximum when k_1 and δ are zero. The expression $2EL_{21}/L_1$ may be regarded as giving the limiting value of the maximum potential at make, to which the actual value would approximate if k_1 and δ were indefinitely reduced. In actual cases, however, k_1 and δ are probably never so small as to allow the above limit to be very closely approached.

Equation (69) shows how V_2 depends upon the constants of the circuits and how these ought to be adjusted in order to diminish as much as possible the secondary potential at make. Thus, V_2 at make is directly proportional to the battery E.M.F. E, and can be diminished by using a battery of lower voltage. In order to maintain the value of the primary current i_0 at break (and the secondary potential at break which is proportional to i_0) the resistance of the primary circuit should also be reduced, and such reduction of resistance would have no serious effect in increasing the potential at make, since by (69) V_2 depends only indirectly (i.e. through k_1 and δ) on the resistance of the primary circuit. In practice, therefore, it is desirable, in order to diminish negative discharge at make, to use as small a battery E.M.F. as is sufficient to produce the required primary current and secondary potential at break.

Again, according to (69) the secondary potential at make is directly proportional to L_{21}, the mutual inductance of the primary and secondary coils. In many coils L_{21} cannot be varied, but in some the primary coil and the core can be moved axially from their symmetrical position in the secondary, a process which diminishes L_{21} without altering L_1. Such a displacement of the primary coil will, therefore, have the effect of diminishing the secondary potential at make and, as we have seen in Chapter II, it may also increase the secondary potential at break.

Another way of reducing V_2 at make is to increase L_1, the self-inductance of the primary circuit, which occurs in the denominator of the expression (69). In Chapter II we have emphasized the importance, from the point of view of the

Break   Make

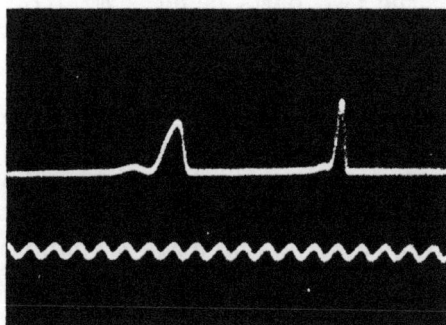

FIG. 52. OSCILLOGRAM SHOWING HIGH SECONDARY
POTENTIAL AT MAKE

action of a coil at break, of increasing L_1 up to a certain point, and we now find that this procedure has the additional advantage of diminishing V_2 at make. The oscillogram shown in Fig. 31, page 75, taken with series inductance in the primary circuit in which the make-potential is scarcely noticeable, illustrates the advantage to be gained in this way.

In some coils, the primary is divided into two or more sections which can be connected in series or in parallel. The change from series to parallel diminishes L_1 and therefore, by (69), increases the secondary potential at make. The oscillogram in Fig. 52 shows the secondary potential (or rather its square) of a coil the two primary sections of which were connected in parallel. In this experiment, the battery E.M.F. was 98 volts, and a rather large non-inductive resistance was included in the primary circuit in order to cut down the potential at break. The potential at make was greater than that at

break, but it should be remarked that the primary resistance
could have been much reduced, in order to increase the break-
potential, without appreciably increasing the potential at
make.

It may be explained here, *apropos* of the sections of the
primary coil, that *for a given P.D. at the primary terminals*
the secondary potential at *break* is also greatest when the sections are connected in parallel. This results, of course, from
the large value of the primary current. Considering, for example, a coil having four equal primary sections, the connection being changed from series to parallel, L_1 and R_1 are both

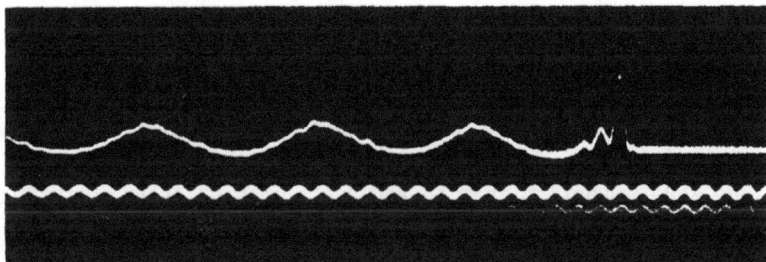

FIG. 53. OSCILLOGRAM SHOWING HIGH SECONDARY POTENTIAL OF
TRANSFORMER WHEN SWITCHED INTO A.C. CIRCUIT

reduced to one-sixteenth, L_{21} to one-quarter, of their "series"
values. Thus, the primary current at break—given approxi-
mately by $\dfrac{E}{R_1}$ $(1 - e^{-\delta T})$, where $T =$ time of contact—is increased to 16 times, and $L_{21}i_0$ to 4 times, its original value.
The coupling being unchanged, the primary capacity being supposed in each case to be the optimum, and the effect of damping
being regarded as unaltered, the secondary potential is proportional to $L_{21}i_0$, and is therefore quadrupled. On the other
hand, if the coil is driven through a considerable series regulating resistance, the spark length at break is greatest when
the layers are in series. In this case the lowering of resistance
due to the change from series to parallel is of little consequence,
and the diminution of L_{21} reduces the secondary potential.

It appears remarkable that the secondary capacity, the value
of which greatly affects the break-potential, has, according to
(69), only an indirect influence on the secondary potential at

make. If this capacity is increased, e.g. by connecting a condenser to the secondary terminals, the period of the oscillation, and with it the time of rise to maximum potential, is lengthened. The damping forces have, therefore, a longer time in which to act, and a reduction of the maximum potential is to be expected. It is only by altering the effect of the damping that the secondary capacity influences the make-potential, and this alteration is but slight if k_1 and δ are small, that is, if R_1/L_1 is small, greater if R_1/L_1 is large.

All the above theoretical conclusions with regard to the make-potential can be easily verified by spark length observations with an induction coil, and it seems probable that in most cases the methods here suggested may be sufficient, without the use of rectifiers, to reduce the reverse current at make to such a small value as to have no deleterious effect.

Another example of a high secondary potential at make is illustrated in Fig. 53, which shows the effect of switching a closed core transformer into an A.C. circuit. It is clear that the potential rises to a considerably higher value in the transient wave at make than it does when the system has subsequently settled down into the alternating current *régime*.

CHAPTER V

THE theory explained and illustrated in Chapters II and III is applicable only to the case of an induction coil, the secondary terminals of which are insulated and, therefore, have no discharge passing between them. In some applications, e.g. in the testing of the dielectric strength of insulating materials, this theory covers the whole of the phenomena up to the point at which the dielectric breaks down. In many cases, however, it is necessary to know how the secondary potential varies during, or is affected by, a discharge. For example, it is well-known that the sparking plug of an engine is very liable to defective insulation owing to the deposition of products of combustion on the surfaces of insulators, and it is important to know how the maximum potential at the spark electrodes is modified by such leakage.

The theory can be extended to cover the case of a coil with secondary discharge if the discharge current is related in some simple manner to the potential at the secondary terminals, e.g. if the current flows in accordance with Ohm's law. In other cases, if the relation between the potential and the current in the discharge is unknown, one must rely entirely upon experiment to discover how the potential is modified by the discharge, and how the discharge current depends upon the constants of the coil and the current supplied to its primary circuit. It may be mentioned here that, during the rapidly varying conditions which succeed "break," the current in the secondary coil (i.e. in its central winding) is, in general, not equal to the discharge current, the difference between them being the current which is absorbed by the capacity of the coil.

Discharge Through a High Non-inductive Resistance. We will first consider the theory of an induction coil with discharge through a conductor in which the current obeys Ohm's law, but which is without self-inductance. The resistance of the conductor being R_3, the current in it, i_3, is V_2/R_3, the wave of discharge current is of the same form as that of the potential V_2 at the secondary terminals, and the discharge current i_3 can be at once evaluated when V_2 and R_3 are known.

Equations (15) and (16) apply without change, viz.—

$$L_1 \frac{di_1}{dt} + L_{12}\frac{di_2}{dt} + R_1 i_1 + V_1 = 0 \quad . \qquad . \qquad . \quad (70)$$

$$L_2 \frac{di_2}{dt} + L_{21}\frac{di_1}{dt} + R_2 i_2 + V_2 = 0. \quad . \qquad . \qquad . \quad (71)$$

V_1 representing, as before, the excess of the potential of the primary condenser over the battery E.M.F. In addition, we have the equations—

$$i_1 = C_1 \frac{dV_1}{dt} \quad . \qquad . \qquad . \qquad . \qquad . \quad (72)$$

$$i_2 = C_2 \frac{dV_2}{dt} + i_3 \quad . \qquad . \qquad . \qquad . \quad (73)$$

$$V_2 = R_3 i_3 \quad . \qquad . \qquad . \qquad . \qquad . \quad (74)$$

Eliminating i_1, i_2, i_3, we have the two equations for V_1, V_2,

$$L_1 C_1 \frac{d^2V_1}{dt^2} + L_{12}C_2 \frac{d^2V_2}{dt^2} + R_1 C_1 \frac{dV_1}{dt} + \frac{L_{12}}{R_3}\frac{dV_2}{dt} + V_1 = 0 \quad . \quad (75)$$

$$L_2 C_2 \frac{d^2V_2}{dt^2} + L_{21}C_1 \frac{d^2V_1}{dt^2} + \left(\frac{L_2}{R_3} + R_2 C_2\right)\frac{dV_2}{dt} + \frac{R_2+R_3}{R_3}V_2 = 0 \quad (76)$$

Assuming $V_1 = Ae^{zt}$, $V_2 = Be^{zt}$, and eliminating the ratio A/B, we find, after reduction, the equation for z

$$(1-k^2)z^4 + \left(\frac{R_1}{L_1} + \frac{R_2}{L_2} + \frac{1-k^2}{R_3 C_2}\right)z^3$$

$$+ \left\{\frac{1}{L_1 C_1} + \frac{1}{L_2 C_2} + \frac{1}{R_3 C_2}\left(\frac{R_1}{L_1} + \frac{R_2}{L_2}\right) + \frac{R_1 R_2}{L_1 L_2}\right\}z^2$$

$$+ \left\{\frac{R_1}{L_1}\cdot\frac{1}{L_2 C_2} + \frac{R_2}{L_2}\cdot\frac{1}{L_1 C_1} + \frac{1}{R_3 C_2}\left(\frac{1}{L_1 C_1} + \frac{R_1 R_2}{L_1 L_2}\right)\right\}z$$

$$+ \frac{1}{L_1 C_1 . L_2 C_2}\frac{R_2+R_3}{R_3} = 0 \quad . \qquad . \qquad . \quad (77)$$

The roots of (77) being denoted by z_1, z_2, z_3, z_4, the solutions for V_1 and V_2 are

$$\left. \begin{array}{l} V_1 = A_1 e^{z_1 t} + A_2 e^{z_2 t} + A_3 e^{z_3 t} + A_4 e^{z_4 t} \\ V_2 = B_1 e^{z_1 t} + B_2 e^{z_2 t} + B_3 e^{z_3 t} + B_4 e^{z_4 t} \end{array} \right\} \quad . \quad (78)$$

The initial conditions are (break occurring at $t = 0$),

$$V_1 = -E, \; V_2 = 0, \; \frac{dV_1}{dt} = \frac{i_0}{C_1}, \; \frac{dV_2}{dt} = 0. \qquad . \qquad (79)$$

Substituting from (78) in (75) and (76), eliminating the A's and using (79), we find the four equations for the B's—

$$B_1 + B_2 + B_3 + B_4 = 0$$

$$B_1 z_1 + B_2 z_2 + B_3 z_3 + B_4 z_4 = 0$$

$$\frac{B_1}{z_1} + \frac{B_2}{z_2} + \frac{B_3}{z_3} + \frac{B_4}{z_4} = G_1$$

$$\frac{B_1}{z_1^2} + \frac{B_2}{z_2^2} + \frac{B_3}{z_3^2} + \frac{B_4}{z_4^2} = G_2.$$

Solving these four equations, we have

$$B_1 = z_1^2 \frac{G_1(z_2 z_3 + z_2 z_4 + z_3 z_4) - G_2 z_2 z_3 z_4}{(z_1 - z_2)(z_1 - z_3)(z_1 - z_4)}, \qquad . \qquad (80)$$

with similar expressions for B_2, B_3, B_4, where

$$G_1 = -\frac{R_3}{R_2 + R_3} L_{21} i_0,$$

$$G_2 = \frac{R_3 L_{21}}{R_2 + R_3} \left\{ C_1 E + \frac{R_3}{R_2 + R_3} \left(\frac{L_2}{R_3} + R_2 C_2 \right) i_0 \right\}.$$

When the resistance R_3 is very large (i.e. several megohms) all four roots of (77) are imaginary and they may then be expressed in the form—

$$z_1 = -k_1 + 2\pi n_1 i$$
$$z_2 = -k_1 - 2\pi n_1 i$$
$$z_3 = -k_2 + 2\pi n_2 i$$
$$z_4 = -k_2 - 2\pi n_2 i,$$

where n_1, n_2 are the frequencies, k_1, k_2 the damping factors of the two component oscillations. The expression for V_2 is then of the form—

$$V_2 = Be^{-k_1 t} \sin(2\pi n_1 t - \delta_1) - B'e^{-k_2 t} \sin(2\pi n_2 t - \delta_2) \quad . \quad (81)$$

If R_3 is moderately large, two of the roots of (77) may be real, and in this case the wave-form is represented by—

$$V_2 = B_1 e^{-\lambda_1 t} + B_2 e^{-\lambda_2 t} - Be^{-kt} \sin(2\pi nt - \delta), \quad . \quad . \quad (82)$$

the real roots of (77) being $-\lambda_1$ and $-\lambda_2$.

When the external resistance is so small that the influence of the capacity of the secondary coil may be neglected, let R_2 represent the total effective resistance of the secondary circuit. In this case the current i_2 is uniformly distributed along the secondary wire, and $L_{12} = L_{21} = M$, the mutual inductance. The equations for V_1 and i_2 are

$$L_1 C_1 \frac{d^2 V_1}{dt^2} + M \frac{di_2}{dt} + R_1 C_1 \frac{dV_1}{dt} + V_1 = 0$$

$$L_2 \frac{di_2}{dt} + MC_1 \frac{d^2 V_1}{dt^2} + R_2 i_2 = 0.$$

The solution is

$$V_1 = A_1 e^{z_1 t} + A_2 e^{z_2 t} + A_3 e^{z_3 t}$$
$$i_2 = B_1 e^{z_1 t} + B_2 e^{z_2 t} + B_3 e^{z_3 t},$$

where z_1, z_2, z_3 are the roots of

$$(1 - k^2)z^3 + \left(\frac{R_1}{L_1} + \frac{R_2}{L_2}\right)z^2 + \left(\frac{1}{L_1 C_1} + \frac{R_1 R_2}{L_1 L_2}\right)z + \frac{R_2}{L_2 . L_1 C_1} = 0,$$

an equation which may be derived from (77) either by making R_3 small or by making C_2 infinite.

The initial conditions being $V_1 = -E$, $i_2 = 0$, $\frac{dV_1}{dt} = \frac{i_0}{C_1}$, we find the following equations for the B's—

$$B_1 + B_2 + B_3 = 0$$

$$\frac{B_1}{z_1} + \frac{B_2}{z_2} + \frac{B_3}{z_3} = -\frac{Mi_0}{R_2} = G_1, \text{ say,}$$

$$\frac{B_1}{z_1^2} + \frac{B_2}{z_2^2} + \frac{B_3}{z_3^2} = \frac{M}{R_2}\left(C_1 E + \frac{L_2 i_0}{R_2}\right) = G_2.$$

The solution of these equations is

$$B_1 = - z_1{}^2 \frac{G_1(z_2 + z_3) - G_2 z_2 z_3}{(z_1 - z_2)(z_1 - z_3)},$$

with similar expressions for B_2 and B_3.

When two of the roots of the cubic for z are imaginary the solution for i_2 takes the form

$$i_2 = Be^{-k_1 t} \sin(2\pi nt - \delta) + B_3 e^{-\lambda t}. \qquad . \qquad . \quad (83)$$

The wave-form of the secondary current thus consists of one oscillatory and one aperiodic component.

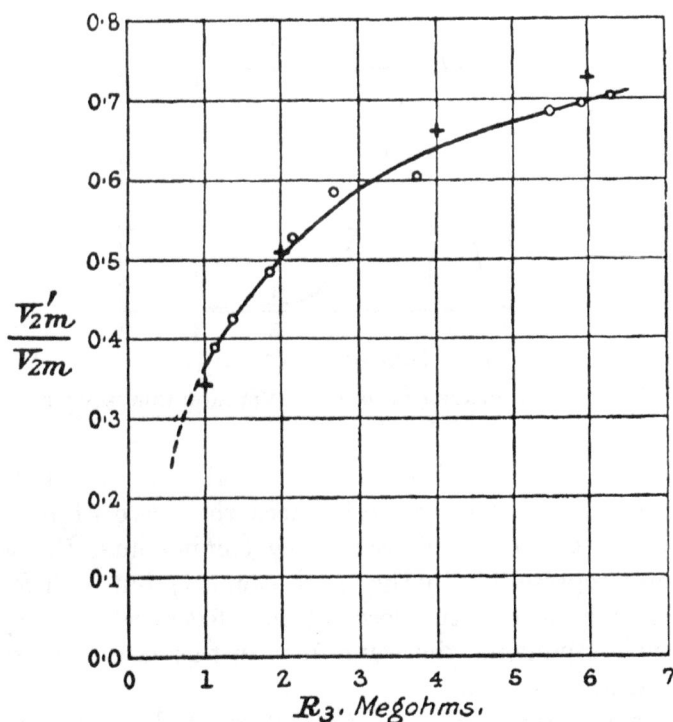

FIG. 54. RATIO OF MAXIMUM SECONDARY VOLTAGE WITH LEAK TO MAXIMUM WITH TERMINALS INSULATED
+ Calculated.   ○ Observed.

In illustration of the foregoing formulae for the effect of a discharge current a number of numerical cases have been worked out for the system, the constants of which are given on page 86, and the wave of secondary potential of which, with secondary terminals insulated (i.e. $R_3 = \infty$) is shown in Fig. 45, page 87. The maximum secondary potential V_{2m}' was calculated by equations (80) and (81) or (82), with $i_0 = 10$ amp., for each of the four values 6, 4, 2, and 1 megohm of the discharge resistance R_3. The ratios of these maxima to the calculated maximum V_{2m} with $R_3 = \infty$ are shown by the four points marked $+$ in the diagram of Fig. 54. It will be seen from the diagram that theoretically the effect of a discharge resistance of 6 megohms is to lower the maximum potential by 27 per cent, that of a resistance of 1 megohm to lower it

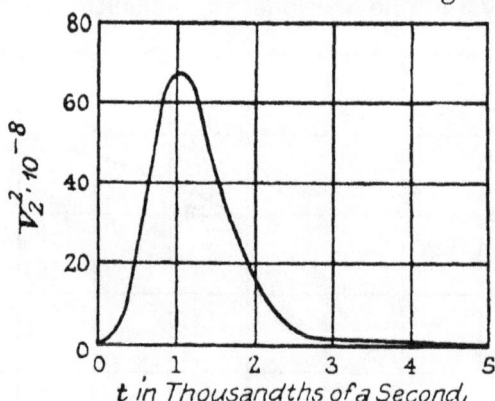

FIG. 55. CALCULATED SECONDARY VOLTAGE CURVE WITH 1 MEGOHM LEAK

by about 65 per cent. The calculations also showed that with the three larger discharge resistances the wave of potential consisted of two damped oscillatory components, the waveform being represented by the expression (81), but with $R_3 = 1$ megohm one of the two oscillations degenerated into two aperiodic components, the wave-form in this case being of the kind indicated by (82).

On the experimental side, the effect of discharge on the maximum secondary potential can easily be examined, either by spark length measurements or with the help of the electrostatic oscillograph. The points marked $\circ$ in Fig. 54 were obtained by the latter method with various high resistances, in the form of tubes containing water, connected between the secondary terminals. The ordinates of these points in the diagram represent the ratios of the maximum potentials observed with, to those observed without, the various resistances R_3 indicated along the horizontal axis. It will be seen that the points $+$ and $\circ$ lie closely on the same curve, indicating satisfactory agreement between the calculated and the experimental results.

With regard to the wave-forms, Fig. 55 shows the calculated $(V_2{}^2, t)$ curve for the case $R_3 = 1$ megohm. It indicates a maximum potential of 82,460 volts at $t = 0\cdot00105$ sec. (the maximum with secondary terminals insulated being 238,700 volts), and the prolonged descent to zero from the peak shows

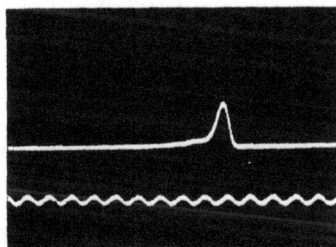

FIG. 56. OBSERVED SECONDARY VOLTAGE CURVE WITH
1·14 MEGOHM LEAK

that the main component of the wave is non-oscillatory. There is still a positive potential of 1,700 volts at $0\cdot005$ sec. The same feature is also shown in the oscillogram of Fig. 56 taken with the electrostatic oscillograph, the discharge resistance R_3 in this experiment being 1·14 megohm.

Fig. 57 shows the calculated curve for the case $R_3 = 2$ megohms. This curve evidently represents a strongly damped oscillation, the potential falling to zero at about $0\cdot0029$ sec., and the slight elevation beyond this point representing the negative half-wave. A similar type of curve is shown in Fig. 58, the oscillogram obtained with a resistance of 2·13 megohms between the secondary terminals. In these curves the more rapid oscillation is too strongly damped to be in evidence; at the peak of the curve in Fig. 57, for example, its amplitude is only 400 volts. The smaller oscillation becomes more marked as the resistance R_3 is increased. In Fig. 59, taken at $R_3 = 5\cdot9$ megohms its effect is just noticeable near the peak of the principal wave. When the water resistance is removed altogether

($R_3 = \infty$) the wave-form, as already stated, is that shown in Fig. 45.

It may be mentioned here that when the secondary terminals of an induction coil are connected by a conductor of considerable resistance, the influence of the capacity of the coil on

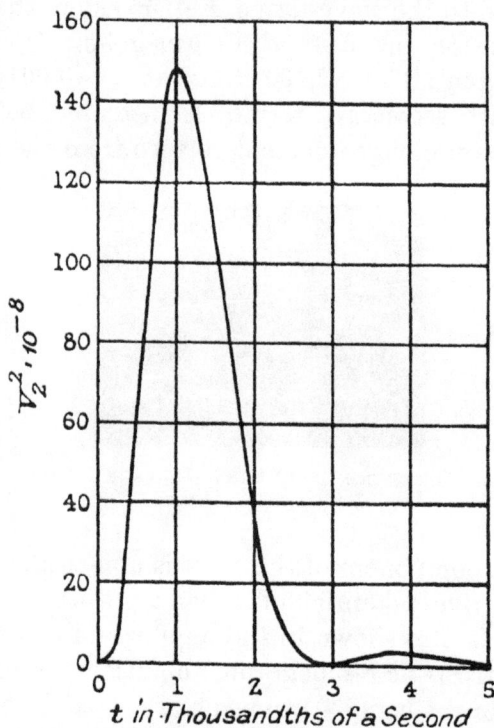

FIG. 57. CALCULATED SECONDARY VOLTAGE CURVE
WITH 2 MEGOHM LEAK

the terminal potential and the discharge current is by no means a negligible quantity. Considering, for example, the case $R_3 = 1$ megohm, i.e. $R_3 = 2\cdot33$ times the effective resistance of the secondary coil, if we had neglected the secondary capacity in our calculation—in which case the coil-current would have been equal to the discharge current, and the wave-form would have been composed of one oscillatory and one exponential component (see equation (83), page 115)—the maximum potential, for the same values of the other constants and a primary current of 10 amp., would have been 126,000 instead of 82,460 volts, a difference of over 50 per cent. The maximum discharge current would have been altered in the same ratio, viz. 126 instead of 82·46 milliamp. The influence of the secondary capacity is still greater at higher discharge resistances. It is clear, therefore, that very great errors can be introduced by neglecting the capacity of the secondary coil in such calculations.

The total quantity of electricity discharged through a given external resistance is, however, not influenced by the capacity of the coil, being equal to the total change of induction through

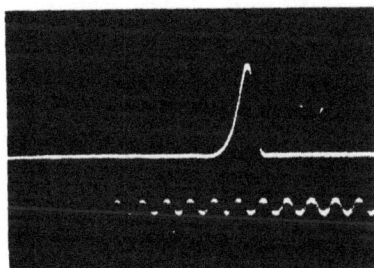

FIG. 58. OBSERVED SECONDARY
VOLTAGE CURVE WITH 2·13
MEGOHM LEAK

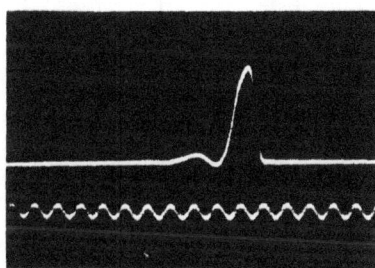

FIG. 59. SECONDARY VOLTAGE
WITH 5·9 MEGOHM LEAK

the secondary at break, $L_{21}i_0$, divided by the total secondary resistance $R_2 + R_3$, where R_2 is the steady-current resistance of the secondary coil. This follows from equations (71), (73), (74), page 112, on integrating over the whole duration of the transient currents at break, and remembering that i_3, V_2, and dV_2/dt, are zero at the beginning and at the end of the period of integration. The water resistances used in the above experiments were measured by observing the throw at break of the moving coil of a ballistic galvanometer—a moving coil milliampere meter provided with a suitable scale and observed through a telescope was used for the purpose—connected in series in the secondary circuit. The instrument was standardized by observations of the throw when the secondary coil was short-circuited through the instrument ($R_3 = 0$). The same method may be used for determining what may be called the "equivalent resistance" of any discharge path, such as a spark gap or an exhausted tube, provided the time occupied by the discharge is short in comparison with the period of swing of the moving coil. The equivalent resistance of such a discharge path means the resistance of a conductor obeying Ohm's law which, if substituted for the given path, would

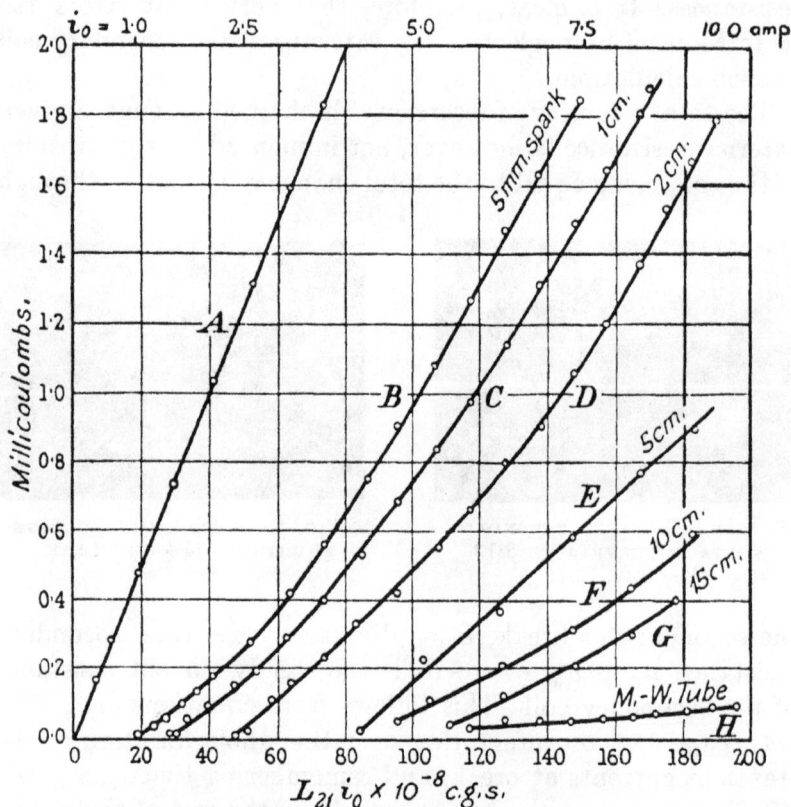

FIG. 60. QUANTITIES DISCHARGED IN SPARKS

allow the same quantity of electricity to be discharged through it for the same change of induction through the secondary coil at break. This change of induction being $L_{21}i_0$, and the resistance of the secondary coil being R_2, the equivalent resistance of the discharge path is—

$$\int \frac{L_{21}i_0}{i_3 dt} - R_2.$$

Spark Discharge. The curves in Fig. 60 represent the results of a series of measurements made in this way of the quantity of electricity discharged in sparks of various lengths between sphere electrodes 2 cm. in diameter. In this diagram the horizontal axis represents the change of induction $L_{21}i_0$ through the secondary at break, the ordinate the number of milli-coulombs discharged. Values of the primary current at break are indicated above the diagram. Points on the straight line A were obtained when the secondary was short-circuited. The ratio of the abscissa to the ordinate of any point on this line is the resistance of the secondary coil at the time when the observations were made, viz. 40,000 ohms. The curves B, C, . . . G, represent discharges through sparks of various lengths, from 5 mm. to 15 cm., as indicated in the diagram. The ratio of the abscissa to the ordinate at any point on one of these curves represents the equivalent resistance of the whole secondary circuit in the circumstances of the experiment. For example, the discharge through the 10 cm. gap at a change of induction of $175 \cdot 5 \times 10^8$ c.g.s. ($i_0 = 9$ amp.) was $0 \cdot 53$ milli-coulomb. The equivalent resistance of the secondary circuit was therefore $175,500/0 \cdot 53$ ohms, i.e. about 331,000 ohms. The equivalent resistance of the gap alone was therefore 291,000 ohms.

Curve H refers to an X-ray tube, the discharge through which, at the same change of induction, was $0 \cdot 076$ milli-coulomb. In this case the equivalent resistance of the secondary circuit was $2 \cdot 31$ megohms, which gives $2 \cdot 27$ megohms as the equivalent resistance of the tube. The equivalent resistance defined in this way, both for spark gaps and for ordinary X-ray tubes, diminishes as the quantity discharged increases ; over the range shown in Fig. 60, most of the curves are concave towards the vertical axis.

Oscillations During Spark Discharge.* The ordinary induction coil spark, when no condenser is connected with the secondary terminals, usually consists of two parts, viz. (1) a single bright initial spark, followed by (2) a series of much less luminous discharges all in the same direction.† This may be seen by examining the spark in a rotating mirror. Thus, in Fig. 61, a photograph of a 10 cm. spark, between platinum points, obtained in this way, the initial spark is followed by

* *The Electrician*, p. 167, 15th Aug., 1919.
† B. Walter, *Ann. d. Phys.*, 66, p. 636, 1898.

a series of broad and much fainter bands representing what may be called a pulsating arc.* The current in this arc goes through maxima and minima without changing sign.

The frequency of the pulsations in the arc is inversely proportional to the square root of the primary capacity, and is, in fact, represented approximately by the expression—

$$\frac{1}{2\pi\sqrt{\{L_1 C_1 (1-k^2)\}}} \quad . \quad . \quad . \quad . \quad (84)$$

obtained by putting $C_2 = \infty$ in the expression (13), page 6,

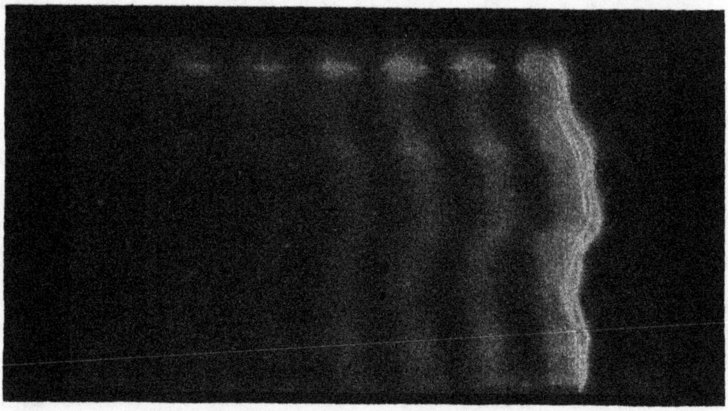

FIG. 61. ORDINARY INDUCTION COIL SPARK WITH
PULSATING ARC

for the frequencies of the system. In other words, it is the frequency of the primary circuit with the secondary terminals connected through a small or moderate resistance. It is probable that by far the greater portion of the quantity of electricity that escapes in this type of discharge passes, not in the initial spark, but in the arc (see the calculation on page 119).

The pulsations in the discharge may also be examined by connecting in series with the spark gap a short spectrum tube containing air at reduced pressure, and observing the narrow part of the tube in the rotating mirror. Fig. 62 is a reproduced photograph of a similar tube when connected in series with a 2·79 cm. spark between sphere electrodes. The bright bands in this illustration correspond to the current pulsations in the

* See, however, a paper by J. Thomson on the nomenclature of the various types of discharge through gases (*Phil. Mag.*, April, 1932).

arc, their sudden cessation after the tenth pulse indicating that the potential at the gap terminals then became insufficient to maintain the arc. In this method the unidirectional character of the pulses is very clearly shown by the blue negative glow appearing at only one end of the tube.

It may be remarked here that a spectrum tube may also be used to show the two component oscillations of a coil which has its secondary terminals insulated (i.e. without discharge).

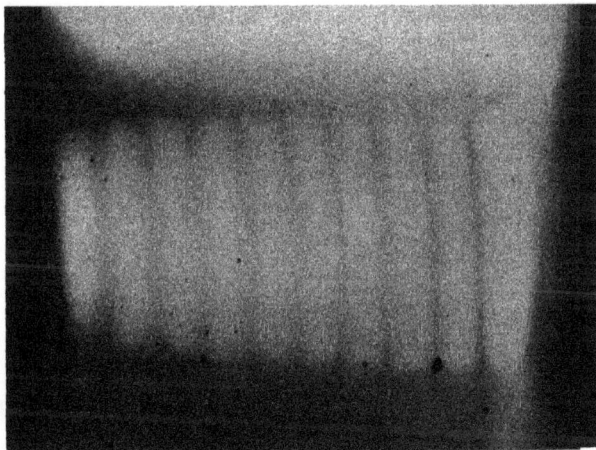

FIG. 62. PULSATING CURRENT THROUGH TUBE IN
SERIES WITH SPARK

For this purpose the tube should be connected between one of the secondary terminals and an insulated conductor of considerable capacity, e.g. one plate of a condenser, and the current through the tube is then the capacity current flowing into the conductor. The photograph reproduced in **Fig. 63** was taken in this way, and it shows clearly the two superposed damped oscillations of the system. The current through the tube in this experiment is alternating, as shown by the cathode glow appearing in turn at both ends of the tube.

In the ordinary spark discharge (with given spark length) the number of pulses in the arc increases with the primary current. Thus, in the experiment of Fig. 61, the primary current at break was about twice, in that of Fig. 62, 2·5 times, the least current required to produce the spark. For a given primary

current the number of pulses in the arc increases as the spark
length is diminished, and when the gap is only a few milli-
metres wide, and the primary current is largely in excess of
the minimum, the bands become finally merged into one un-
broken band of diminishing intensity representing a continuous
or aperiodic arc.

All these observations on the arc portion of the ordinary
discharge correspond, qualitatively at least, with the solution
represented by equation (83), page 115, for the case in which
the secondary terminals are connected through a resistance

Fɪɢ. 63. Dᴏᴜʙʟᴇ Oꜱᴄɪʟʟᴀᴛɪᴏɴ Sʜᴏᴡɴ ʙʏ Tᴜʙᴇ ɪɴ Sᴇʀɪᴇꜱ
ᴡɪᴛʜ Cᴏɴᴅᴇɴꜱᴇʀ

which is not very great. The expression (83) represents a discharge current consisting of an oscillatory superposed upon an
exponentially decaying component.

The view of the matter suggested by these observations is
(1) that the initial spark represents the escape of the static
electricity accumulated on and near the secondary terminals
at the moment when the discharge begins; (2) that the secondary current (beginning with the value which it has at this
moment) flows through the arc as an aperiodic decaying current; and (3) that superposed upon this aperiodic current is
an oscillatory current having the period of the system. It is
clear that, on this view, the arc should become alternating
if the secondary current immediately before the discharge
begins is sufficiently reduced, and this is found to be the
case. In the oscillations preceding the discharge the secondary current is zero at the moment of maximum potential

$(i_2 = C_2 \dfrac{dV_2}{dt} = 0)$, so that if the primary current is adjusted to a value equal to, or only slightly exceeding, the minimum required to produce the spark, the discharge begins with very little current flowing in the secondary coil. Two or three pulses are then generally observed, and these are found to be alternating, i.e. a negative band has appeared between the first two positive bands. This effect is shown in Fig. 64, a reproduced photograph of the spectrum tube when in series with

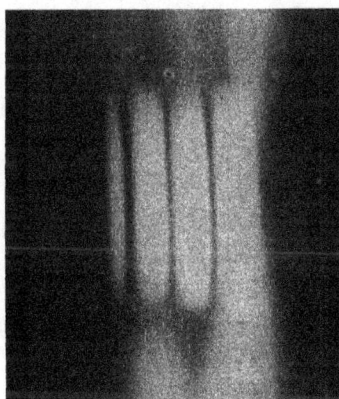

FIG. 64. ALTERNATING BANDS IN SPARK DISCHARGE

a 6·5 cm. spark between sphere electrodes, the primary current only slightly exceeding the minimum required to produce the spark. The alternating character of the discharge is here shown chiefly by the manner in which the luminosity of the narrow part of the tube is connected with that surrounding the electrodes, this connection being much more marked at the negative than at the positive end. On the ground glass plate it is also shown by the difference of colour, the cathode glow appearing alternately at the upper and lower ends of the tube. The alternating bands of Fig. 64 are clearly distinguished from those of Fig. 63 by the facts that in the experiment of Fig. 63 there was no discharge and that in the photograph reproduced in Fig. 64 there is only one oscillation. In fact, the bands in Fig. 63 represent capacity current, those in Fig. 64 represent discharge current.

As the primary current is increased, the second or reversed band of Fig. 64 becomes narrower and ultimately disappears. A trace of a narrow reversed pulse is seen between the first two positive pulses of Fig. 61, where it is shown only by the glow at the negative electrode.

Still another way of showing the pulsating current in the ordinary discharge is by means of the current oscillograph, this instrument being connected in the secondary circuit in series with the spark. Fig. 65 shows two curves obtained in

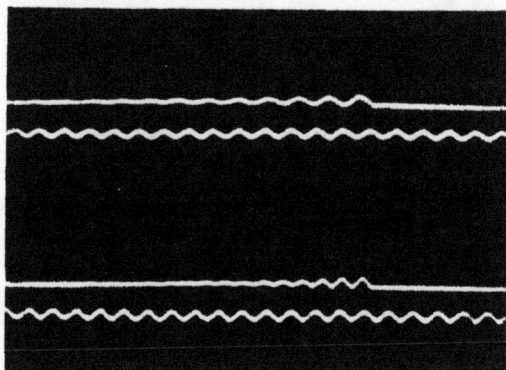

FIG. 65. CURRENT OSCILLOGRAMS OF SPARK DISCHARGE

this way, each of which indicates the small unidirectional pulsating current in the arc. In the upper curve, the primary capacity was 2 mfd., in the lower 1 mfd., and the frequencies of the two waves are in the inverse ratio of the square roots of these numbers. Probably, the initial spark contains a high-frequency oscillation, but this would, of course, not show in a curve obtained with a mirror oscillograph.

With regard to the variation of potential in the ordinary spark discharge, the potential first rises, as explained in Chapters II and III, to the value required to produce the spark, and then falls with great rapidity to zero or to a value comparable with the low potentials which prevail during the arc portion of the discharge. The two potential curves in Figs. 66 and 67 (for 18·4 and 12·1 cm. sparks respectively) show the rapid fall of potential to zero in the initial spark, but on the scale of these illustrations the low potential during the arc would not be noticeable. In these two experiments, the primary current was only slightly in excess of the minimum required to produce the spark, the arc was, therefore, of very short duration, and the small oscillations seen in Figs. 66 and 67

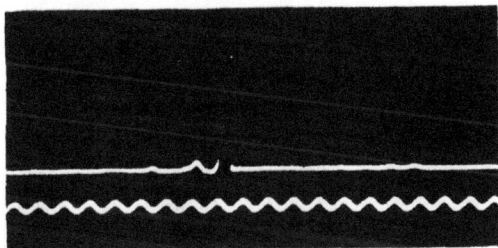

Fig. 66. Potential Oscillogram of 18·4 cm. Spark Discharge

after the rapid fall to zero are due to the energy remaining in the coil system after the discharge has ceased.

It will be seen in Fig. 61 that the arc bands have the same shape as the initial spark, showing that the arc and spark currents travel along the same path. It is evident that in this type of discharge, in which the initial spark represents the discharge of a very small capacity, this spark leaves its path sufficiently ionized to allow the subsequent arc current to

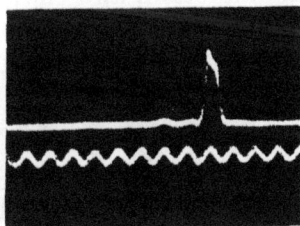

Fig. 67. Potential Oscillogram of 12·1 cm. Spark Discharge

traverse the gap at a comparatively very low potential. The property of a spark of leaving its path ionized seems, however, to depend very greatly upon the capacity associated with the secondary coil.

Spark Discharge with Secondary Condenser. When the spark electrodes are connected with the plates of a condenser, and the primary current is moderately in excess of the minimum required to produce the spark, the discharge, instead of consisting of a bright spark followed by a faint pulsating arc, now takes the form of a series of sparks. Two examples of this type of discharge are given in Figs. 68 and 69, each showing

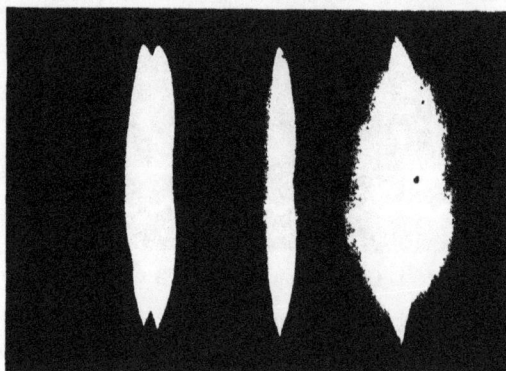

FIG. 68. MULTIPLE SPARK DISCHARGE OF CONDENSER

the discharge between platinum points at one interruption of the primary current, the plates of a glass condenser being connected with the spark terminals. The secondary capacity was the same in both experiments, but the primary capacities were different. The current in the successive sparks in these illustrations is in the same direction, and they follow one another

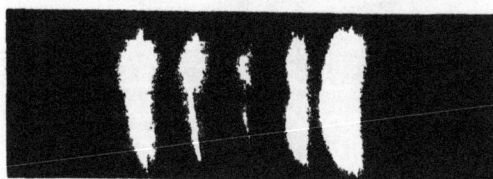

FIG. 69. MULTIPLE SPARK DISCHARGE OF CONDENSER

with a certain regularity, the interval between them being directly proportional to the square root of the capacity connected across the interrupter. The frequency of the discharges is, in fact, given approximately by the expression (84), page 122, for that of the primary circuit with the secondary short-circuited, being but slightly, if at all, affected by the capacity connected with the secondary terminals.

The potential is usually highest in the first and the last of the sparks which are also usually the brightest. The sparks frequently become multiple, especially if they are rather short, so that they appear in the rotating mirror as a number of groups. Thus, in Fig. 68, the third spark is double, indicating a tendency on the part of the condenser to exhibit a number of partial discharges rather than one complete discharge. With a narrower gap, the number of sparks in a group may be considerably greater, as in the photograph reproduced in Fig. 70, taken with a 4 mm. spark in parallel with an oil condenser. If the primary condenser is disconnected, the grouping disappears, as shown in Fig. 71. The frequency of the sparks in a group is increased by reducing the capacity connected with the spark

FIG. 70. GROUPED MULTIPLE
SPARKS

FIG. 71. UNGROUPED
MULTIPLE SPARKS

terminals, is diminished by increasing the spark length, and does not seem to be affected by varying the secondary self-inductance, e.g. by drawing out the primary and core along the axis of the secondary. The frequency is, of course, much less than that of the high-frequency oscillations of the circuit formed by the condenser, its connecting wires, and the spark gap; each spark, presumably, contains a train of these oscillations.

The total number of sparks occurring in a single discharge of the induction coil, i.e. at a single interruption of the primary current, increases with the primary current up to a certain point, when it suddenly becomes reduced to a small number, sometimes only one or two. These are then seen to be followed by a very faint pulsating arc. At this transient stage the discharge may take the form either of—

1. A large number of sparks (as many as 60 were found in one photograph),* or of

2. A small number of sparks followed by a faint pulsating arc, the arc frequently showing a few sparks at its end as well as at its beginning.

* Examples of short multiple sparks are shown in Fig. 111, p. 211.

The first type of discharge is the more suitable for exciting a Tesla coil, or other high-frequency oscillation transformer, from the secondary of which fuller and brighter sparks—they are also evidently multiple—can be drawn when the exciting discharge is of the form (1) than when it is (2). As the primary current i_0 is increased the arc becomes brighter, and the appearance in the rotating mirror becomes similar to that of Fig. 61, the period of the arc pulses also being given by the same expression.

It appears from these observations that the presence of a condenser in parallel with the spark gap has a considerable influence in hindering the formation of an arc, though it does not altogether prevent it. The ionization which a condenser spark leaves in its path is not, unless the primary current greatly exceeds the minimum required to produce the spark, sufficient to prevent subsequent high potential sparks from appearing, but is sufficient to make the secondary coil behave (as shown by the frequency of the grouped sparks) as if its terminals were connected by a moderately small resistance. The tendency of the condenser to show partial discharges is less easy to explain, but probably all the phenomena are associated with the strong high-frequency oscillating current flowing in the local circuit formed by the condenser and the spark gap.

The difference between the ordinary and the multiple spark types of discharge is of much importance in connection with the theory of spark ignition, which will be considered in Chapter VIII.

Discharge Through an X-ray Tube.* It has already been indicated that the quantity of electricity passing in the discharge through a high vacuum is much less, for a given change of induction through the secondary coil, than that which passes in the ordinary spark discharge in air at atmospheric pressure (see curve H, Fig. 60, page 120). The discharge through X-ray tubes is further illustrated by the curves in Fig. 72, which show the quantity, in microcoulombs, passing through two X-ray tubes at a single interruption of various primary currents. The corresponding changes of induction are

* *The Electrician*, p. 168, 15th Aug.; p. 201, 22nd Aug., 1919. *Journ Rönt. Soc.*, 16th April, 1920.

given below the diagram of Fig. 60. At the time when the measurements of Fig. 72 were made, the two tubes were in a rather "soft" (low vacuum) state, and the form of the curves shows that the quantity discharged increases more rapidly than in proportion to the change of induction through the secondary circuit. Apart from the fact that the curves do not begin at zero—no discharge passes unless the primary current exceeds

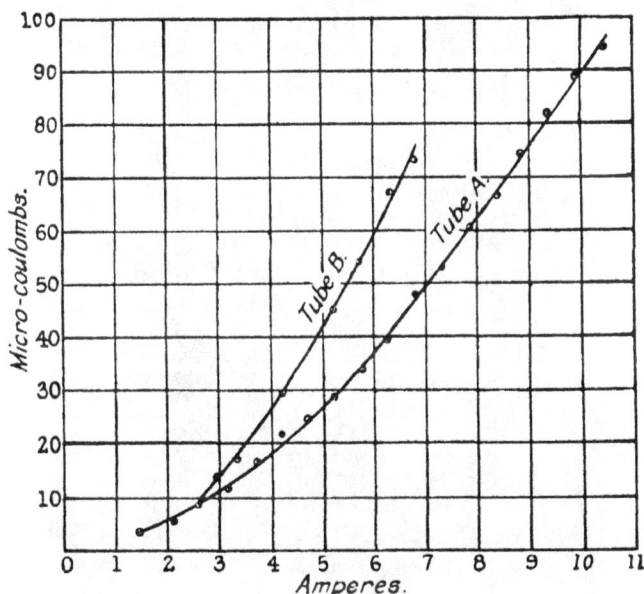

FIG. 72. QUANTITIES DISCHARGED THROUGH X-RAY TUBES

a certain minimum value—the deviation from proportionality is, however, within the range covered by Fig. 72, not so great as to lead one to expect that the wave-form of potential during the discharge would differ very greatly from that found when the secondary terminals are connected through an ordinary high resistance. For soft X-ray tubes the wave-form is, in fact, more or less of this type, that is, the potential falls rather gradually, though somewhat irregularly, from the maximum value, with considerable prolongation of the time occupied by the first half wave of the principal oscillation. Some examples of these wave-forms, taken with the electrostatic oscillograph, are shown in Fig. 73.

The three curves in Fig. 73 were obtained at the interruption of the same primary current (10·35 amp.), and when the tube was disconnected the wave-form was that shown in Fig. 45, page 87. In the uppermost curve, the maximum potential is

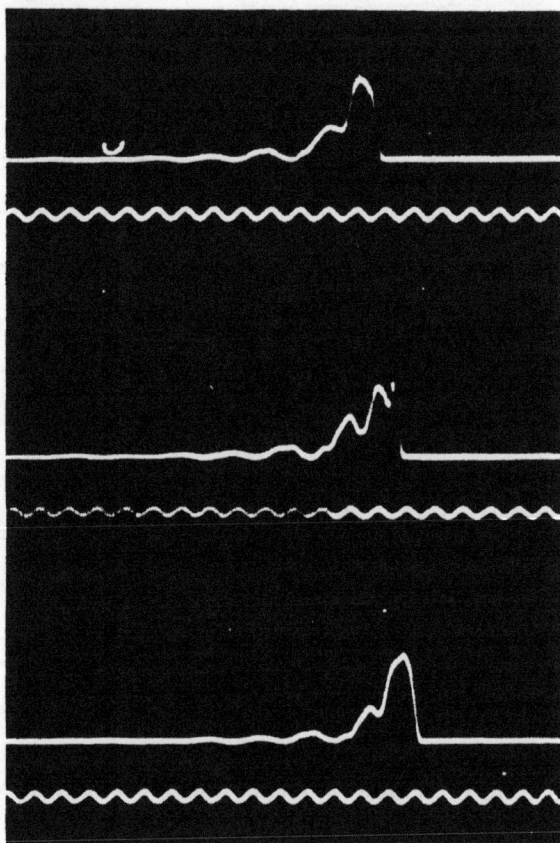

Fig. 73. Potential Oscillograms of X-ray Tube

96,000 volts, and to judge by the rounded form of the curve near the summit the discharge begins before the maximum is reached, probably at 85,000 volts. From the maximum, the potential falls gradually to a minimum at 57,000, and then rises to a second maximum of 60,000 volts from which it falls to a value of about 30,000 to 35,000 volts at which the discharge ceases. After this, the potential varies with the usual two frequencies of the coil with open secondary. The whole duration of the discharge is about 0·0028 sec., which is much less than that of the ordinary spark discharge in air at atmospheric pressure.

If we take the mean potential during the discharge in Fig. 73 as 70,000 volts, we can form an estimate of the energy spent in the discharge. From curve *A* (Fig. 72) we find that the quantity of electricity passing through this tube at $i_0 = 10·35$ amperes is 94 microcoulombs. Thus, the energy dissipated in the discharge is $70,000 \times 94 \times 10^{-6} = 6·6$ joules. The initial energy in the primary circuit $\frac{1}{2}L_1 i_0^2$ being 9·6 joules, the ratio

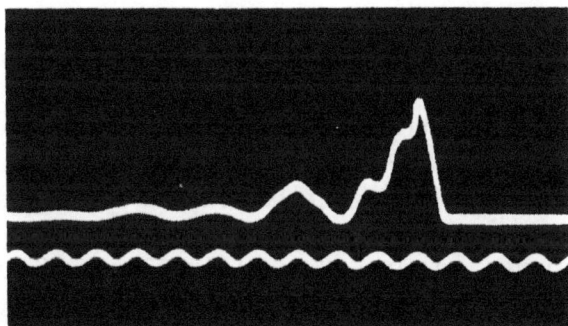

FIG. 74. POTENTIAL OSCILLOGRAM OF X-RAY TUBE

of the energy spent in the tube to the energy supplied to the system, i.e. the "efficiency of discharge," is about 69 per cent. Probably only a small fraction of the energy of discharge is, however, usefully employed in generating X-rays.

The two lower curves of Fig. 73 are of the same general type, but there are differences in detail. The maximum potential is greatest in the lowest curve (98,000 volts), and there is a pronounced oscillation in the descending portion of the second curve. The frequency of this oscillation is about 760 per sec., which corresponds to that of the more rapid component of the system with the secondary terminals connected through a resistance of rather over 2 megohms. When another vacuum tube was connected in series with the X-ray tube, the frequency of the oscillation on the descending portion of the curve was increased, as shown in Fig. 74. In this case, the frequency is about 960, which corresponds to the more rapid component when the terminals are connected to a resistance of nearly

4 megohms. If the primary current is much increased, the oscillation becomes more marked, and a long train of these oscillations may sometimes be observed as the potential falls from its maximum value.*

An examination of these wave-forms, therefore, leads us to the conclusion that in some respects the discharge through a soft X-ray tube follows a similar course to that of the discharge through a moderately high ohmic resistance, but that there is a remarkable difference between the two cases in regard to the damping of the two component oscillations of the system. If the same coil were used to produce discharge through an ordinary resistance of 2 or 4 megohms, both components would be oscillatory, and both strongly damped, but the more rapid component would be so much more strongly damped than the other that its influence on the wave-form would be scarcely noticeable (see Figs. 57, 58, 59). On the other hand, during the discharge through a soft tube, the slower component may be so strongly damped as to become aperiodic, while the more rapid component remains strongly in evidence. The full explanation of this difference has not been discovered, but the present experiments seem to suggest that the strong damping of the slower component in a gas tube is due to that portion of the current which is carried by positive ions, while the rapid fluctuations of the current representing the higher frequency component are chiefly conveyed by the electrons. It is, at any rate, certain, as we shall see later, that the cathode ray current does fluctuate in accordance with the frequency of the more rapid component of the system. In a metallic conductor, it is believed, the whole current is conveyed by electrons, but their free path in a metal is so short that collisions are frequent and both component oscillations are strongly damped in accordance with Ohm's law.

When the tube had been hardened by running for some time with a motor interrupter, the potential curve was of a quite different character. Instead of showing the aperiodic fall of potential from the maximum, the curve for the hard tube had much more closely the general form observed when the secondary

* The oscillations are well shown by an oscilloscope tube connected in series with the X-ray tube.

terminals were insulated, but the first half-wave contained a number of deep indentations, indicating that a portion of the discharge took place in a number of sharp pulses, each accompanied by a rapid fall of potential. An example of this type of curve, showing three pulses, is given in Fig. 75. The number of pulses increases with the primary current and their frequency, which is quite unrelated to any of the frequencies of the coil system, may be several thousands per second at strong currents. Similar sharp indentations in the potential curve for

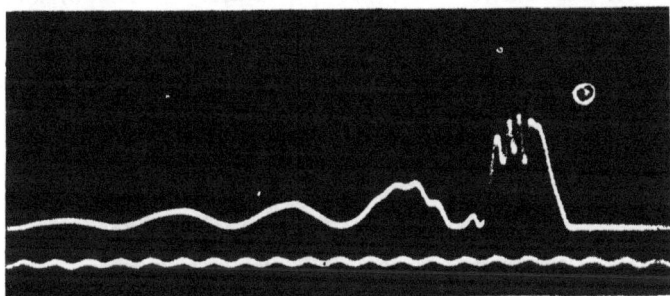

FIG. 75. POTENTIAL OSCILLOGRAM OF HARD TUBE

hard tubes have been observed by A. Wehnelt,* who experimented with a coil supplied with alternating currents.

It seems clear that the current pulses represented by these indentations in the potential curve are occasioned by the nature of the discharge path rather than by any property of the system employed in generating the potential. Their relation to the discharge current does not appear to have been definitely ascertained, but it is certain that they do not represent the whole of the discharge current through the tube, and that they are accompanied by similar fluctuations in the cathode ray portion of the current. This latter fact may be demonstrated by photographing the X-rays emitted by the tube (through a narrow slit in a lead screen) simultaneously with the potential wave-form on a moving plate.† It will be found that the X-ray band on the plate shows a number of intermittences corresponding exactly with the indentations of the potential curve. The fact that the pulses represent only a small

* *Ann d. Phys.*, 47, p. 1112 (1915).
† See *Theory of the Induction Coil*, pp. 158–60.

proportion of the whole discharge current may be shown by examining a spectrum tube, connected in series with the X-ray tube, in a rotating mirror. The spectrum tube shows an intense discharge band without any trace of fluctuations corresponding to the pulses.

Further experiments are desirable to ascertain the exact nature and cause of these current pulses which appear to be characteristic of the discharge through gases within a certain range of low pressure.

Delayed Discharge. Another peculiarity often observed in the discharge through high vacua is the very variable potential

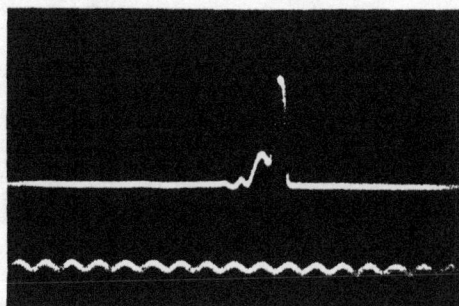

FIG. 76. DELAYED DISCHARGE

at which the discharge begins. The discharge through a tube in a fairly "soft" condition may normally begin at a comparatively low potential, but sometimes the commencement is delayed until the potential has risen to a much higher value. An example of this delayed discharge is shown in Fig. 76, in which the potential rises nearly to the maximum attainable with the tube disconnected before falling rapidly to the normal value at which the later portions of the discharge take place. The potential curve for the normal discharge through the same tube is shown in Fig. 77. The normal and the delayed discharge may occur at the same value of the primary current, and when this is the case, the abnormal discharge is indicated by a much less intense fluorescence of the tube. The abnormal discharge may also be indicated by a spark gap connected in parallel with the tube, the spark length being of course greater when the discharge is delayed.

In some cases, the commencement of the discharge is so long delayed that it does not occur at all in the first half-wave, but takes place in the second, or negative, half-wave of the potential oscillation. When this happens, the discharge through

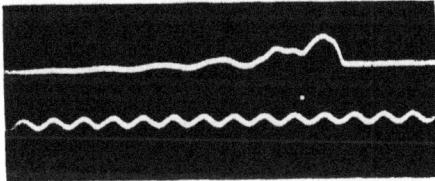

FIG. 77. NORMAL DISCHARGE

the tube is in the wrong direction, as is clearly shown by the form of the fluorescent area of the tube, or by a ballistic galvanometer connected in series with the tube, which gives a deflexion in the negative direction. This reversal of the discharge at break was observed by Duddell* when examining

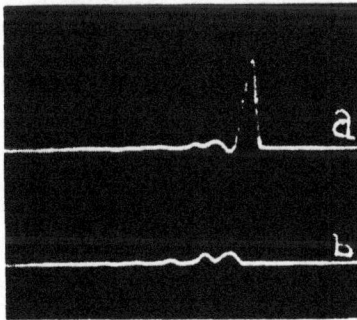

FIG. 78

a = Reversed discharge at break.   *b* = Direct discharge at break.

the current through an X-ray tube by means of an oscillograph connected in series with it.

The reversed current at break is illustrated by the two potential oscillograms of Fig. 78, the upper of which shows the potential wave-form when the discharge is delayed to the second half-wave and is, therefore, negative, the lower curve representing the normal, or direct, discharge. The corresponding wave-form when the tube was disconnected was similar to that

* *Journ. Rönt. Soc.*, iv, 17 (1908).

shown in Fig. 48. The two curves of Fig. 78 were taken at
the same primary current, but the discharge can always be
made to pass the right way by sufficiently increasing the current. The maximum secondary potential attainable without
discharge appears to be the chief factor determining the direction of the discharge. The higher this maximum is the less
likelihood is there of the discharge being delayed beyond it
and becoming reversed.

Discharge Through a Coolidge Tube. It might be expected
that when the discharge takes the form of a pure electron

FIG. 79. POTENTIAL OSCILLOGRAM OF COOLIDGE TUBE

current, as in a very high vacuum tube with hot cathode, the
wave-form would be of a rather simpler type than those obtained in experiments with gas tubes, and this anticipation
is confirmed by the following observations on a Coolidge tube.
In Fig. 79 is shown the wave-form of the terminal potential
of a Coolidge tube when the filament current was 4·5 amp.—
nearly the maximum value for this tube. When the filament
current was zero, the curve was similar to that of Fig. 45,
page 87. As the filament current is increased from zero (with
the primary current kept constant), the amplitude of the curve
diminishes, that of the second and following half-waves becoming smaller relatively to the first, showing that an increasing proportion of the energy of the system is being absorbed
by the discharge in the first half-wave. At the same time the
form of the first half-wave varies, the second peak, initially
considerably higher than the first, becoming relatively smaller.
This effect is to be expected as a consequence of the increased
damping of the oscillations arising from the growing discharge

through the tube. The same change is observed, for example, when a water resistance connected with the secondary terminals is reduced in value.

When the filament current reaches 4·0 amp., the two peaks of the first half-wave are equal, at 4·25 amp. the second peak is decidedly smaller than the first, and by this time the second and succeeding main half-waves have practically disappeared. At stronger filament currents the more rapid oscillation still persists, and it is well marked in Fig. 79. If the primary capacity is now increased, the depression in the curve representing the more rapid component moves back along the curve, as it does when the secondary terminals are insulated, indicating an increasing frequency-ratio. The two peaks become equal again at about 5 mfd.

The curve of Fig. 79 shows no indication of the sharp indentations observed in the case of an ordinary high-vacuum X-ray tube, nor is there any evidence of a prolongation of the time occupied by the positive half-wave as is found with a rather low water resistance and in an ordinary low-vacuum tube. The time of return to zero potential is, in fact, apparently shorter in Fig. 79 than when the filament current is zero.

The maximum potential indicated in Fig. 79 is about one-half the maximum attained at the same primary current with the filament cold. The same proportional reduction of maximum voltage would be produced by connecting a water resistance of about 2 megohms between the secondary terminals (see Fig. 54, page 115). The Coolidge tube curve of Fig. 79 does, in fact, resemble the 2-megohm curves of Figs. 57 and 58, with one important difference, however, viz. that in the discharge through an ohmic resistance of 2 megohms the more rapid oscillation is very much more strongly damped than the slower component and makes no visible appearance in the oscillogram, while in the Coolidge tube curve showing the same proportional diminution of maximum potential, and, therefore, indicating the same damping of the slower component, the small oscillation is still strongly in evidence. The result seems to confirm the suggestion, already made in the discussion of the observations on gas tubes (see previous paragraph), that in the discharge through gases at low pressure, the electron

current favours the more rapid component oscillation of the
system.

If we compare the Coolidge tube curve of Fig. 79 with that
of a gas tube of fairly low vacuum (see Fig. 73) we find here a
greater difference. The gas tube absorbs much more strongly
the energy of the slower component, with the result that this
component easily becomes aperiodic, and has its first half-wave
considerably prolonged. This effect seems to be completely
absent in the Coolidge tube, at least it is so within the range
of the present observations. As to the more rapid component
this is treated much in the same way by the two kinds of tube,
neither having any marked damping effect upon it. It may be

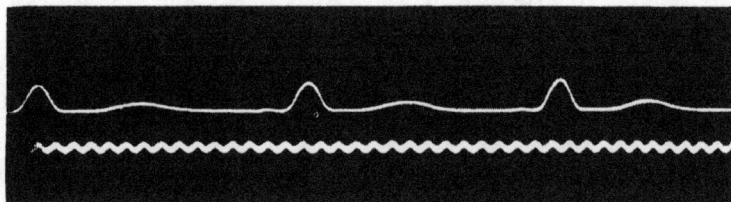

FIG. 80. POTENTIAL OSCILLOGRAM OF COOLIDGE TUBE ON A.C.
CIRCUIT

said, in fact, that a Coolidge tube deals with the slower component approximately as an ohmic resistance would do, but
in its treatment of the more rapid oscillation, it more closely
resembles the gas tube.

Another marked difference between the curves of Fig. 73
and that of Fig. 79 is that in the former the oscillations of
the system continue after the discharge has ceased, while in
the latter there is no trace of such oscillations after the first
positive half-wave. This difference doubtless arises from the
fact that in the gas tube the discharge begins and stops at
a finite potential, so that there is a considerable amount of
energy left in the system (then subject to comparatively small
damping forces) when the discharge has run its course. In the
Coolidge tube on the other hand, the discharge begins from zero
and continues until the potential again becomes zero, so that
the energy can, if the filament emission is sufficiently great,
be entirely absorbed in the first half-wave.

Fig. 80 is a curve obtained with the electrostatic oscillograph

connected to the terminals of a Coolidge tube excited by a
high-tension transformer supplied with alternating current.
The higher peaks in the curve represent, of course, the negative
portions of the wave, during which no discharge passes. In
the three smaller half-waves the potential is much reduced
by the (rectified) discharge.

On X-ray Production. The question of the most effective
adjustment of a coil for X-ray production does not appear
to have received very much attention, although it is desirable
in several investigations and applications of X-rays to be able
to produce the most intense beam possible with the generating
apparatus available. The following is an account of an experiment on this subject made by the writer, in which a comparison
is made of some adjustments, but which is far from being an
exhaustive inquiry into the subject. The method adopted was
to expose one-half of a photographic plate to the rays produced in one adjustment of the coil, the other half to the rays
in another adjustment, the number of breaks and the primary
current at each break being the same for both halves. The
plate was wrapped in thick paper and placed inside a cardboard box to the interior of which was attached an oblong
metal plate composed of five parts. One end of the plate was
of aluminium 1 mm. thick, covered with lead 1·27 mm. thick,
the next section of aluminium 1 mm. thick, the remaining three
sections in order, of aluminium 0·75 mm., 0·5 mm., and 0·25 mm.
thick respectively. Thus, the photographic effect of the rays
produced in any two adjustments could be compared after the
rays had passed through the cardboard and paper alone, or after
they had passed through the above thicknesses of aluminium.

The box containing the plate was mounted in a certain
position 2 ft. from the centre of the tube. A lead screen could
be moved over the box so as to cover either half of the plate,
and at the same time to cover either half of the composite
metal plate. Preliminary experiments showed that the slight
difference of position of the two halves of the plate did not
cause any difference in the photographic effect upon them.
No attempt was made to measure the density of the photographic negatives, the object being merely to find the "optimum" of a number of adjustments.

The coil used in the experiment had its four primary sections all connected in parallel, and the first series of observations had for their object the determination of the optimum primary capacity. In the X-ray exposures, 200 flashes were allowed to fall upon each half of the plate, the primary current (10·0 amp.) being adjusted after each 25 breaks. Comparison of the effects of various capacities taken in pairs showed that 3 mfd. was better than 2 mfd. and than smaller values, and that 4 mfd. was better than 5 and larger capacities. Compared directly, 3 and 4 mfd. gave equal effects, the two halves of the plate being equally dense. Thus we may take 3·5 mfd. as the best capacity for X-ray production in the circumstances of these experiments. With the tube connected but with a primary current insufficient to cause the tube to glow, the longest spark appeared when C_1 was about 2 mfd. At a primary current of 10 amp.—which current was about 3·3 times the least current required to cause the tube to glow—the most effective primary capacity for X-ray production was thus about 1·75 times the capacity which gave the highest secondary potential when the terminals were insulated.

The system was now varied by adding series inductance, in the form of an air-core coil of about 0·005 henry, to the primary circuit, the coupling being thus reduced to about 0·54. In this case the capacity which gave the highest secondary potential with secondary terminals insulated was 3·5 mfd. At 10 amp. the most effective capacity for X-ray production was found, in the manner just described, to be 4·5 mfd. It appears, therefore, that if the primary current at break is not more than 3 or 4 times as great as the least required to produce discharge, the most suitable capacity for radiographic purposes is not much greater (1·3 or 1·75 times as great in the present experiments) than the value which gives the highest potential with the terminals insulated.

The "optima" in the above two cases were then compared directly, i.e. one-half of the plate was exposed to the rays when the coil was provided with the series inductance and 4·5 mfd., the other half with 3·5 mfd. and no series inductance. The exposure of each half was 200 flashes, the current at each break 10·0 amp. The result showed a very marked difference in favour of the series inductance. This photograph is shown in Fig. 81, the lighter portions representing those parts of the plate on which fell the more intense photographic rays. The lowest (and largest) section of the composite plate is the lead-covered portion.

It will be seen that the rays produced when the series inductance was in circuit (left-hand half) were much more intense photographically than those emitted when the inductance was

absent, both before and after their passage through the various thicknesses of aluminium. A comparison of the two wave-forms in this experiment showed that the maximum potential at the terminals of the tube was 1·25 times as great with the series inductance as without it. It is, however, not always the case that maximum potential is accompanied by maximum intensity of radiation (compare, for example, the effects of delayed discharge, page 136).

It is, of course, to be expected that the energy of the discharge will be greater with than without the series inductance in these experiments, since the primary current at break was constant, and therefore more energy was supplied to the sys-

FIG. 81. COMPARISON OF X-RAY EFFECTS WITH DIFFERENT PRIMARY INDUCTANCES
Equal primary currents.

tem when the extra coil was in circuit; for a given primary current, the energy supplied, represented by $\frac{1}{2}L_1 i_0{}^2$ is proportional to the total primary self-inductance. It does not follow from the experiments just described that the efficiency of X-ray production is improved by inserting external inductance in the primary circuit. This point was tested by making the comparison when the quantities of energy supplied to the system were equal. The ratio of the primary self-inductances with and without the extra coil was, in the present case, 1·43. Consequently, if the current with the additional coil in circuit be $1\sqrt{1\cdot43}$ times the current when the air-core coil is disconnected, the energies supplied will be equal. Accordingly, one-half of the plate was exposed at 10·0 amp., 3·5 mfd., and with no series inductance, the other half at 8·38 amp., 4·5 mfd., and with the air-core coil.* The photograph is shown reproduced in Fig. 82, the right-hand half being the exposure with the series coil in circuit. It again indicates a decided superiority of the radiographic effect with the additional primary inductance. It follows, therefore, that in the circumstances of the present experiments not only is the photographic effect of the X radiation, for a given primary current, improved by adding external series inductance to the primary circuit, but the efficiency of production of the rays is also increased by this means.

FIG. 82. COMPARISON OF X-RAY EFFECTS WITH DIFFERENT PRIMARY INDUCTANCES
Equal energies in primary.

When the primary current at break is increased, the (radio-graphically) most effective primary capacity also becomes greater. The improvement effected by the use of extra inductance coils in the primary circuit was observed at much stronger currents, also when the primary sections were connected in series. It is an advantage from the practical point of view that the optimum primary capacity for X-ray production increases with the primary current, since with stronger currents a larger capacity is in any case required to ensure satisfactory working of the interrupter.

A much more extended series of experiments on X-ray production was carried out by Professor J. A. Crowther,† who photographed simultaneously on a moving plate the waveform of the terminal potential of an X-ray tube, that of the current through the tube, and the beam of X-rays transmitted through a narrow slit, a portion of the beam passing also through an aluminium wedge. The chief types of potential wave-form were obtained, and in particular, Professor Crowther showed that when the wave-form of a soft gas tube consists

* The energy supplied was, in fact, rather less with than without the air-core coil, since, owing to the variable permeability of the iron core, the self-inductance of the primary coil was smaller at 8·38 than at 10·0 amperes.
† *Brit. Journ. Radiol*, XX, April (1924); XXI, April (1925).

of an aperiodic component with a long train of oscillations superposed upon it (see page 133), the X-ray impression on the plate is also drawn out into a long band with intermittences corresponding to the oscillations of the system. This shows clearly that (as indicated in the previous paragraph) the cathode ray current forms at least a considerable part of the rapidly fluctuating component of the current through the tube.

Among the conclusions drawn by Professor Crowther from his experiments are—

1. That the intensity of the X radiation is proportional at each instant to the product of the values of the current and the square of the potential.

2. That in working conditions the voltage on a Coolidge tube is far beyond the saturation value.

3. That the peak current through a gas tube is proportional to the difference of the squares of V, the peak potential on the tube, and V_0, the breakdown voltage of the tube.

4. The square of the peak voltage on a gas tube is approximately proportional to the primary current at break.

Recently, a method of X-ray production has come into favour in which a condenser, previously charged to a high potential, is discharged through the tube. The writer is not aware that this method has yet been made the subject of oscillograph study; it would be of much interest to compare the wave-forms obtained in this method of exciting the tube with those observed when coils or transformers are employed.

CHAPTER VI

THE DIFFRACTION OF ELECTRONS BY THIN FILMS*

IN this chapter we will give an account of some experiments on the diffraction effects exhibited when a narrow pencil of cathode rays is transmitted at high velocity through a thin film of solid material. With a view to indicating the nature of the effects to be expected, it will be well to preface the description of the experiments with a short account of the theoretical aspect of the subject.

Matter and Radiation. One of the most remarkable developments of modern research in physics is concerned with the relations between matter and radiation. The more closely the properties of radiation (that is, heat, light, X-rays, etc.) are studied, the greater the number of them which appear to be identical with those possessed by ordinary material bodies. The idea of the similarity between radiation and matter may be said to have originated in 1873, when Clerk Maxwell† showed that, according to his electromagnetic theory of light, a beam of light should exert a pressure upon the surface of a body on which it falls. The magnitude of this pressure was shown to be equal, if the incidence is normal and if the radiation is absorbed by the surface, to the energy contained in unit volume of the beam. The pressure of light was demonstrated experimentally by Lebedef in 1899, and a few years later it was studied by the late Professor Poynting who introduced the idea that since light exerts thrust on a surface upon which it is incident it may be regarded as possessing momentum. If we consider a beam of unit cross-section incident normally upon the surface, the energy in unit volume of the beam being E, and the velocity of light being c, the pressure E on the surface is equal to the momentum destroyed by the surface per second, that is, the momentum contained in a length c

---

* A considerable part of the substance of this chapter was contained in the first Selby lecture, given at University College, Cardiff, on 3rd March, 1931. See also *Phil. Mag.*, p. 641, Sept. (1931).

† *Treatise on Electricity and Magnetism*, Vol. II. Sect. 792.

of the beam. Consequently, the momentum in unit volume of the beam is equal to E/c. Now, whatever be the nature of light, it is something that travels with velocity c, and this must be one factor in the expression for the momentum. The other factor must be of the nature of mass, and denoting by m the "mass" per unit volume of the beam, we have the relation—

$$mc = \frac{E}{c}$$

or $E = mc^2$. . . . . (85)

So far, we have arrived at a number of points of resemblance between radiation and matter. Radiation possesses energy, momentum, and something akin to mass and density, all well-known properties of ordinary material bodies. These properties may, however, equally well be said to be possessed by other kinds of radiation. Sound waves, for example, exert pressure on a surface, and the momentum per unit volume in a beam of sound may be represented by the energy per unit volume divided by the velocity of propagation of sound. The specially close analogy between light (including heat, etc.) and matter, which distinguishes light in this respect from sound and other kinds of wave motion, is arrived at by other considerations.

By a quite different line of argument, based upon Einstein's Special Theory of Relativity, it has been shown that the energy of any material particle of mass m can be expressed as mc^2, c being the velocity of light. This includes the total internal energy of the particle as well as the ordinary kinetic energy due to its motion. The latter portion is represented by the variation of the mass of the particle with its velocity according to the equation—

$$m = \frac{m_0}{\sqrt{(1 - v^2/c^2)}}$$
. . . . . . (86)

m_0 being the mass of the particle when at rest relatively to the observer.

It appears, therefore, that not only does a beam of light

possess energy, but the expression for the amount of this energy is of precisely the same form as that which represents the energy of a material particle.

Another remarkable resemblance between radiation and matter is derived from thermodynamic considerations. It is found that the radiation in an enclosed vacuous space is subject to the same thermodynamic laws as those applicable to material systems. The radiation possesses energy and, owing to the pressure which it exerts upon the walls it does work when the enclosure expands. The laws of thermodynamics, when applied to such an enclosure at a uniform temperature, are found to lead to results which agree perfectly with experiment. For example, the Stefan-Boltzmann law that the energy of the radiation per unit volume is proportional to the fourth power of the absolute temperature may be deduced by thermo-dynamic reasoning and has been amply verified by experiment. It is true that the relation between pressure and temperature is not the same for radiation as it is, for example, in the case of a gas, but the fundamental thermodynamic laws are precisely the same for both.

A still more striking point of resemblance between radiation and matter has emerged from the quantum theory. In Planck's original form of the quantum hypothesis there was no sugges-tion that radiation possessed anything in the nature of an atomic structure. This idea was introduced later by Einstein, and the view that radiation exists in the form of groups of waves, called variously "light quanta," "light particles," or "photons," the energy of each group being proportional to the frequency of the waves in it, has since received more and more general acceptance owing to its adaptability in the explanation of many of the phenomena of radiation.

When we think of these numerous properties, dynamical, thermodynamic, atomic, which matter and radiation possess in common, it does not seem very unnatural to go a step further and assume that every particle of matter has in some way associated with it a group of waves, the energy of which represents the energy of the particle. This is the starting-point of de Broglie's undulatory theory of matter.*

* *Ondes et Mouvements* (Paris, 1926).

Bringing in the quantum relation we have two equivalent expressions for the energy of the particle, viz.—

$$mc^2 = h\nu \qquad . \qquad . \qquad . \qquad . \qquad . \qquad . \qquad . \qquad (87)$$

where h is Planck's constant, and ν is the frequency of the associated waves

If v is the velocity of the particle, its momentum is—

$$mv = \frac{h\nu}{c^2/v}$$

$$= \frac{h\nu}{u} \qquad . \qquad . \qquad . \qquad . \qquad . \qquad . \qquad (88)$$

where u is a velocity related to the velocity of the particle by the equation

$$uv = c^2 \qquad . \qquad . \qquad . \qquad . \qquad . \qquad . \qquad (89)$$

The velocity u is identified by de Broglie with that of the waves associated with the particle. It is clear that, v being less than c, the phase velocity of these waves must be greater than that of light.

Further, the phase velocity of the waves divided by their frequency, that is, u/ν, is equal to their wavelength. Denoting the wavelength by λ we find from equation (88)—

$$\lambda = \frac{h}{mv} \qquad . \qquad . \qquad . \qquad . \qquad . \qquad . \qquad (90)$$

Equations (89) and (90), giving the velocity and the wavelength of the associated waves, are the fundamental equations of de Broglie's theory.*

By means of equations (86), (89), (90), the wavelength λ can be expressed in terms of the velocity of propagation u, the result being—

$$\lambda = \frac{h}{m_0 c^2}\sqrt{u^2 - c^2} \qquad . \qquad . \qquad . \qquad . \qquad . \qquad (91)$$

The wavelength of the associated waves, therefore, increases with their velocity of propagation. Now, it is a well-known

---

* De Broglie showed, by an application of the Lorentz transformation, that a series of stationary waves associated with a particle at rest would appear, to an observer moving relatively to it with velocity $-v$, as a system of waves the phase of which travels with velocity u, the amplitude with velocity v.

fact in the theory of wave motion that if the velocity of propagation depends upon the wavelength (as in a dispersive medium), the group velocity differs from that of the waves, being in fact equal to $u - \lambda du/d\lambda$. In the present case, the group velocity is easily seen, by (91), to be equal to v.

The view presented by de Broglie's theory is, therefore, that a particle of mass m moving with velocity v is in and moves with (and perhaps consists of) a group of waves, the phase velocity of which is c^2/v, the wavelength of which is h/mv, and the frequency of which is, by (86) and (87)

$$\nu = \frac{m_0 c^2}{h\sqrt{1 - v^2/c^2}} \qquad . \qquad . \qquad . \qquad . \qquad . \qquad . \qquad (92)$$

The frequency of the waves does not vary much with the velocity of the particle unless this velocity approaches that of light.

For example, the waves of an electron travelling at 10^{10} cm. per sec., that is, at one-third of the velocity of light—a speed easily attainable in a cathode ray tube—have, according to (90), a wavelength of $0 \cdot 0685 \cdot 10^{-8}$ cm., which would correspond to the wavelength of very hard X-rays. The frequency of these electron waves would be, by (92), $1 \cdot 309 \cdot 10^{20}$ which is much greater than that of X-rays of the same wavelength, and the waves travel at a speed equal to three times that of light.

It may be noticed that the *charge* of an electron does not enter into these expressions for its wavelength and frequency, but since "mass" and "charge" are apparently inseparable—no particle is known to exist unassociated with charge, no charge free from mass, and according to one view the mass of a particle is entirely due to the charges which it contains—there must be some fundamental relation between them,[*] and there is every probability that if matter can be described in terms of waves, the same may be said of electricity

Apart from these considerations, it has been shown by Sir J. J. Thomson,[†] as a consequence of electromagnetic theory,

---

[*] Apparently, such relation cannot be of the most simple type, for a proton has a mass 1,845 times that of an electron while their charges are numerically the same.

[†] *Phil. Mag.* v., p. 191, 1928.

that a moving charged particle is accompanied, owing to variations in the distribution of the lines of force which it carries with it, by a system of waves the wavelength of which is inversely proportional to the momentum of the particle, and the phase velocity of which is related to the velocity of the particle by equation (89). One important difference between the two theories is that in the theory of de Broglie the product of the wavelength and the momentum is, by equation (90), an absolute constant, whereas in the theory of Sir J. J. Thomson, this product depends also upon the distribution of electrons in the medium through which the waves are passing. It is a matter for experiment to show whether the wavelength is determined solely by the momentum or whether there is any evidence indicating that the wavelength can vary independently of the momentum.

The First Experiments on Electron Diffraction. The first experiments which showed that electrons possess wave characteristics were those of Davisson and Kunsman,* who directed a pencil of cathode rays, emitted by a hot filament and accelerated by various voltages up to 1,000, at 45° on to a metal surface. The electrons scattered by the metal in different directions were collected and measured, and the results were found to show maxima and minima of intensity similar to those observed in optical diffraction experiments.

More definite results were obtained later by Davisson and Germer,† who directed the pencil normally on to the surface of a single crystal of nickel, and examined the rays scattered in different "azimuths" as well as in different "latitudes." They found very distinct maxima of intensity in azimuth, corresponding to various reflecting planes of the crystal, and from their directions and the measured potentials they were able to calculate the wavelength and the velocity of the electrons and so compare their results with the relation expressed by the de Broglie equation (90). On the whole, they found that the quantity $\lambda mv/h$, which by (90) should be unity, did not differ greatly from this value, but that it showed a systematic diminution with increasing voltage. Thus, for one set

---

* *Physical Review*, 22, p. 242 (1923).
† *Nature*, 119, p. 558 (1927).

of reflecting planes $\lambda mv/h$ fell off by about 16 per cent as the potential was increased from 54 to 174 volts. Still the wavelengths measured by Davisson and Germer agreed, so far as their order of magnitude was concerned, with the values indicated by the de Broglie theory.

Another method was first used by Professor G. P. Thomson,* who transmitted the pencil of cathode rays, generated at much higher potentials by an induction coil, through thin solid films and received them on a photographic plate. Films of celluloid, gold, aluminium, platinum, and another substance were used, and the photographs showed systems of concentric diffraction rings, usually uniform in intensity round the circumference but, in some cases, with maxima on certain radii indicating scattering by a single crystal of the substance of the film. By deflecting the transmitted pencils in a magnetic field, Professor Thomson showed that the diffracted rays were electrons and not X-rays. From measurements of the diameters of the rings, spark-gap measurements of the potential on the tube (and hence the velocity of the electrons), and from the known crystal structure of some of the materials examined, Professor Thomson found close agreement with the de Broglie equation (90), without any systematic deviation from it.

The reflexion method and the transmission method have since been studied by several other experimenters, with various sources of potential, various modifications in the apparatus, and with films of different substances. A list of references to some of these papers is given at the end of the present chapter.

The Present Experiments. The following is an account of the present writer's experiments on the subject, the main object of which was to obtain further evidence with regard to the relation between the velocity and the wavelength of electrons. The experiments were designed so as to yield perfectly simultaneous determinations of velocity and wavelength, a condition which is necessary if strictly corresponding values of these quantities are desired.

A diagram of the discharge tube and camera used in the present experiments is shown in Fig. 83. The thick-walled

* *Proc. Roy. Soc.*, A, 117, p. 600 (1927); 119, p. 651(1928); 125, p. 352 (1929).

brass tube B serves to connect the discharge tube to the camera and also acts as anode. At each end of this tube is a circular aperture about 1 mm. in diameter. About halfway along the brass tube C is a hinged shutter, S, which lies when open along the lower part of the tube, but can be closed by a magnet. At c is a very small aperture in a sheet of tinfoil stretched on a brass ring, and at about 2 mm. beyond the tinfoil is the thin film F mounted on a smaller ring which can be placed in different positions relative to the axis of the larger ring so as to

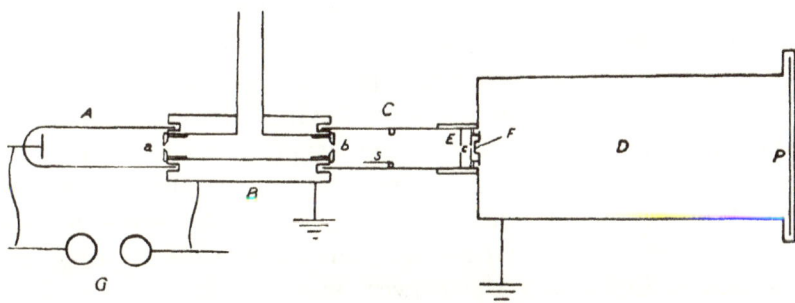

FIG. 83 DISCHARGE TUBE AND CAMERA

A = Discharge tube.
B = Connecting tube with apertures a b.
C = Brass tube with shutter S, diaphragm E, small aperture c, and film F.

D = Camera.
P = Plate.
G = Spark gap.
M = Magnet.

allow different parts of the film to be examined. A diaphragm E cuts off rays which would pass outside the rings into the camera, without hindering the free passage of air from the camera into C. From the side-tube in B a connection (including a liquid-air pocket) leads to a three-stage mercury diffusion pump and to a McLeod gauge.

The principal dimensions are: cathode to a 7 cm.; a to b 10 cm.; b to c 8·5 cm.; film to plate 18·6 cm.

A spark gap G, having zinc sphere electrodes 2 cm. in diameter, is connected directly to the cathode and the anode, from which there are also leads to the secondary terminals of an induction coil.

The three apertures a, b, c are collinear, and small deflexions of the cathode-ray pencil due to local magnetic fields were sufficiently well corrected by bar-magnets placed horizontally as at M or vertically at the side of the apparatus.

One method of procedure adopted in taking a photograph was to exhaust the apparatus to a sufficiently high vacuum, and then to open the shutter (removing the magnet which held it closed), and to pass a single discharge through the tube, using a hand-operated mercury interrupter. The examination of a few plates taken in this way indicated the most suitable positions in which to place the magnets.

In this method the spark gap is opened out wide, so that no spark accompanies the flash which produces the photograph, and the wave of potential applied to the electrodes of the discharge-tube is of the kind in which the potential rises rapidly to its maximum value and descends more gradually from it, the descending portion having a wave superposed upon it corresponding to the oscillation of the induction coil in the circumstances of the experiment. The current through the tube follows a similar course. The cathode-ray stream is, therefore, very heterogeneous, the faster rays starting first (or nearly first), and the slower ones following in the later portions of the discharge. The rings formed have considerable radial width, the inner edge being due to the rays of greatest speed, and the outer portions, formed later, to the more slowly moving rays. This radial separation of the rings for different speeds might cause considerable overlapping when several rings are present, but, chiefly owing to the fact that the rings obtained have maxima of intensity round the circumference, often on different diameters in consecutive rings, there is in many cases no difficulty in identifying the rings in spite of the heterogeneity of the rays. There are also methods, as we shall see later, by which the production of the slower cathode rays in the discharge can be almost entirely prevented, so that only the rings for the maximum speed are formed, or by which the rings due to rays of different speed can be separated out laterally on the plate.

By observing the spark length (the shutter being closed) before and after the exposure a rough idea of the maximum potential during the exposure may be obtained. When greater accuracy was required a rather different method of procedure was adopted.

The Diffraction Patterns. In Fig. 84 is reproduced a photograph taken with this apparatus, showing a ring having a regular variation of intensity round the circumference. This photograph was obtained by the method just described, i.e. by a single flash discharge with the spark gap opened out. In this experiment the film was of celluloid, too thin to show colours, and therefore appearing rather dark, in reflected daylight. The aperture c was oval in shape (as shown by the form of the central spot), having a greatest diameter of 0·25 mm. and a smallest of 0·1 mm.

The ring forms a complete circle, but has twelve maxima

FIGS. 84 AND 85. CELLULOID

of intensity nearly equally spaced on its circumference, indicating that, though there are in the celluloid film reflecting planes of given spacing inclined at a certain angle θ to the direction of the incident rays and otherwise arranged at all angles about this direction, there are six sets of these planes which reflect more strongly than the others.

With a smaller aperture c (circular, 0·05 mm. diameter) a number of other photographs were obtained with celluloid films showing a ring with maxima in the form of spots or radial lines on the circumference, sometimes twelve maxima as in Fig. 84, sometimes twelve arranged as six pairs, and in some cases only six maxima. Fig. 85 is an example showing a ring with six maxima.*

Apparently the smallest number of maxima on a ring is six for celluloid films, and the twelve maxima of Fig. 84 must be due to two neighbouring portions of the film, each having three principal sets of reflecting planes inclined at about 60°

---

* The short "tail" attached to the central spot in Fig. 85 will be referred to on page 161.

to each other, the planes of one portion being inclined at nearly 30° to those of the other.

The photograph reproduced in Fig. 86 was taken with a celluloid film which was the nearest approach to a "black" film that the writer was able to obtain. It was prepared in the usual way by evaporating a very dilute solution of celluloid in amyl acetate on a water surface, and the black portion was found to be separated by a sharp boundary line from the thicker parts of the film adjacent to it. The discharge was a "single flash" produced by the interruption of a primary current of 26 amp., and the maximum potential in the discharge was about 58,000 volts.

It will be noticed that the radial streamers of which the pattern in Fig. 86 is composed are so arranged that their heads, or inner ends, lie mainly on three concentric circles. In the inner circle the six maxima are prominent and broad, and there are other finer maxima, but the circle is not continuous. The maxima of the second ring lie in the angles between the six maxima of the first. In the third ring the positions of the maxima are not so distinct, but from an examination of a number of similar photographs it was concluded that they lie chiefly on the same diameters as those of the first ring. In some of the photographs the third ring was almost continuous, and was surrounded by a fourth ring which was quite continuous. Details as to the ratios of the diameters of the rings are given later.

Photographs like that reproduced in Fig. 86 are quite suitable for the determination of the ratios of the diameters of the rings in a pattern. The inner ends of the radial lines are sharply defined, owing to the fact that the "peak" of the cathode-ray current coincides with that of the potential wave,* and they form good marks for measurement; but when it is required to measure the diameter of any one ring corresponding to a given value of the maximum potential applied to the tube these photographs are not suitable. Owing to the slight spreading of scattered pencils on their way between the film and the plate, the spreading being inwards (radially) as well

---

* When the rings are continuous the inner edge is sharply defined for the same reason, as in Fig. 84.

as outwards, a diameter measured between the extreme inner ends of the radial lines is rather too small. A better plan is to isolate the pencils of maximum velocity and measure between the centres of the spots which they form on the plate.

This may be effected fairly well by setting the spark gap to a potential less than the normal maximum which the coil is capable of giving in the circumstances of the experiment, and allowing the spark to pass simultaneously with the discharge which produces the photograph. The result of this is that when the spark passes the potential falls with extreme rapidity to small values, the current in the usual slowly

FIGS. 86 AND 87. CELLULOID

descending portion of the current wave is all cut out, and the photograph reduces to a number of spots representing very short lengths of the radial lines of Fig. 86. Fig. 87 shows a reproduced photograph taken by this "simultaneous spark" method, with a single discharge accompanied by a 28·0 mm. spark (60,000 volts). Owing to the absence of the slower rays the central spot is smaller and the long radial streamers are not formed. The intensity of the spots in Fig. 87 is also less than that of the corresponding parts of Fig. 86. This is because the time for which the tube electrodes are within a small range of the maximum potential, and therefore the production of cathode rays of this potential, is diminished when the spark is allowed to pass. At lower potentials a single flash with spark was not sufficient to give a good photograph, but by this method the discharges may be repeated without fear of blurring the pattern, since in each discharge the maximum potential has a definite value determined by the spark length. The chief advantage of this method is, of course, that the maximum potential is determined for the actual discharge in which the photograph is taken, and the photograph is formed practically by cathode rays of this potential alone.

Even with this method, however, the cathode rays incident on the film are not quite homogeneous. Many of the spots in the first ring of Fig. 87 have short radial "tails" attached to them, which are probably due to cathode rays generated while the potential on the tube is rising to the sparking value. Some of the spots, however, are small and round, and these were selected as the most suitable for measurement of diameters of diffraction rings. The diameters of the round spots are about 0·2 mm. for an aperture c of diameter 0·05 mm.

FIGS. 88 AND 89. CELLULOID

Fig. 88 and Fig. 89 are two more photographs taken by the "simultaneous spark" method, each with a considerable number of flashes, and the sharpness of the inner edges of the spots in these photographs indicates the degree of exactness with which repetitions of a pattern may be obtained by this method. The film was of celluloid, the aperture c 0·0012 sq. mm. in area, and the first photograph, Fig. 88 (800 flashes with 25 mm. sparks), shows six maxima on each ring, while the other, Fig. 89, taken with a different part of the same film (600 flashes with 23 mm. sparks) has six pairs of maxima on each ring.

Fig. 90 is a reproduced photograph taken with a gold film,* the exposure being 500 flashes with 27·0 mm. sparks. Single flash photographs (without sparks) were obtained with gold, but they were too faint to be suitable for reproduction. The ratios of the diameters of the principal rings in Fig. 90 and the positions of the maxima upon them are characteristic of the

---

* A piece of gold leaf thinned with dilute *aqua regia*.

face-centred cubic lattice, with four edges of the cube in a definite direction parallel to the surface of the film. One ring, a very faint ring smaller than the smallest of the principal rings, may be due to reflexion by (111) planes, indicating that in some part of the film the cubic cells are placed in a different way from those of the main lattice.

In all the photographs obtained with the present apparatus and gold films, the principal rings showed maxima as in Fig. 90; in none were they uniform in intensity round the circumference.

FIG. 90. GOLD

According to the Bragg method of treating the reflexion of X-rays by crystals, each spot or maximum on a diffraction ring, with its diametrically opposite spot, is formed by reflexion of the waves at a certain set of planes in which all the atoms, or scattering centres, of the crystal may be arranged. The diameter of the ring depends upon the "spacing" of the planes, i.e. the normal distance between consecutive planes of the set, and the positions of the maxima on the rings are determined by the orientation of the planes. The diameters of the rings and the positions of the maxima on them therefore guide us in deciding which sets of planes are responsible for the formation of the various rings in a pattern.

It may be well here to explain the notation used in describing the orientation of the various sets of planes of a crystal, and it will be sufficient for our purpose to take the case of a simple cubic lattice as an example. Taking rectangular axes of reference Ox, Oy, Oz, coinciding with three edges of a cubic cell, and the edge of the cell as the unit of length, then the set of planes denoted by (111) is that set which is parallel to the plane intersecting each of the axes at unit distance from the origin. This particular plane is the nearest of the set to the origin without passing through it. Similarly, the set of planes (211) is parallel to the plane which intersects Ox at 1/2, and Oy and Oz each at 1. The set (100) is parallel to the plane which intersects Ox at 1 and does not intersect the other axes at all, i.e. this set is parallel to Oy and Oz. The three numbers in brackets are called the "indices" of the set of planes, and the spacing of the set is equal to the length of the edge of the cubic cell divided by the square root of the sum of the squares of the indices. Thus, the spacing of the (111) planes is $a/\sqrt{3}$, if a is the edge of the cube, and the spacing of the (211) planes is $a/\sqrt{6}$.

In the case of gold, the lattice is not of the simple cubic type, the cubic cell having an atom at the centre of each face as well as at each corner. The lattice is, therefore, described as "face-centred cubic." In this case, the indices of any set of planes are either all even or all odd. Thus, the (111) plane is still the nearest of its set to the origin, but of the planes parallel to Oy and Oz the nearest to the origin is now (200), not (100). In each case the indices denote the reciprocals of the intercepts on the axes of the plane of the set which is at the shortest finite distance from the origin. Thus, if we again take a as the side of the cubic cell, the spacing of the (200) planes of a face-centred cubic lattice is $a/2$, that of the (551) planes is $a/\sqrt{51}$. The maxima on a ring formed by reflexion at (551) planes are on diameters inclined at 45° to those of the (200) ring formed by the same crystal, and the diameter of the (551) ring is greater than that of the (200) ring approximately in the ratio $\sqrt{51}/2$.

These relations are also applicable, as we shall see later, to the reflexion of electron waves by the crystal planes of a gold film through which the waves are transmitted.

Transmission of Dispersed Pencils. In addition to the method of the simultaneous spark there is also another method by which the diffraction rings formed by the cathode rays of maximum speed can be separated from the others. If the incident pencil is deflected by a magnet so as to fall with sufficient obliquity upon the aperture *c*, some of the rays strike against one side of this aperture and are deflected by it (in the opposite way to the deflexion produced by the magnet), and the slower rays are deflected more than those moving more rapidly. Presumably this deflexion is produced by the reflecting planes of the metal screen containing the aperture *c*, as in the original experiments of Davisson.

On the plate, therefore, we find the usual round spot now drawn out into a band, as shown in Fig. 91, a photograph

FIG. 91
DISPERSED PENCIL

FIG. 92
PATTERN FORMED BY
DISPERSED PENCIL

taken without a film. The maxima and minima on the band represent the oscillations of the coil, and the interval between them can be varied by changing the capacity connected across the interrupter. They imprint a kind of time-scale on the photograph, and show that the maximum cathode-ray current occurs very early in the discharge, and that the current afterwards falls off much in the same way as does the total current through the tube.

If such a dispersed pencil is transmitted through a thin film each of the maxima on the band produces its own pattern, so that the rings for different electron velocities are separated from each other laterally. Usually the deflected rings are not complete, or are much stronger on the side towards which they are deflected, so that the first ring formed—i.e. the ring corresponding to the rays of greatest velocity—is left fairly clear and is suitable for measurement. A photograph taken in this way is shown reproduced in Fig. 92, and Fig. 85 is another example.

In some of the illustrations similar to Fig. 86, the radial streamers show clearly the maxima and minima representing the oscillations of the coil, giving the appearance of a large number of concentric rings. In these cases also the rings formed by cathode rays of different speeds are well separated (radially) from each other.

The Velocity of the Electrons, and the Wavelengths. It is generally admitted that the wavelength of electron waves within a film, which determines the dimensions of the diffraction pattern, differs from their wavelength before they enter the film, but as the relation between these wavelengths is still uncertain we shall regard them as quantities to be determined independently from the experimental observations.

If V is the potential difference of the electrodes of the discharge tube the approximate expressions for v, the electron velocity due to the potential V, λ the de Broglie wavelength *in vacuo*, and λ' the wavelength corresponding to a first order diffraction ring of diameter D produced by reflexion at planes of spacing d, are

$$v = \sqrt{\frac{2eV}{m_0}\left(1 - \frac{3}{4}\frac{eV}{m_0 c^2}\right)}. \qquad \qquad (93)$$

$$\lambda = \frac{h}{\sqrt{2em_0}\,V}\left(1 - \frac{1}{4}\frac{eV}{m_0 c^2}\right) \qquad \qquad (94)$$

$$\text{and } \lambda' = \frac{Dd}{2l} \qquad \qquad \qquad (95)$$

l being the distance from film to plate.

The first is derived from the energy equation

$$Ve = (m - m_0)c^2$$

on substituting $m = m_0/\sqrt{1 - v^2/c^2}$, expanding, solving for v and using the approximate value $\sqrt{2eV/m_0}$ for v in the small term. The second is obtained from the de Broglie equation $\lambda = h/mv$, and the third from the Bragg equation $n\lambda = 2d \sin \theta$, with $n = 1$, and 2θ, the angle between a scattered pencil and the primary rays, which is small for the innermost rings in all the present experiments, equal to $D/2l$.

With the numerical values

$$e = 1\cdot591\ .\ 10^{-20}\ \text{e.m.u.,} \quad e/m_0 = 1\cdot769\ .\ 10^7\ \text{e.m.u./gm.,}$$
$$h = 6\cdot56\ .\ 10^{-27}\ \text{erg sec.,} \quad l = 18\cdot6\ \text{cm.,}$$

and V being the potential in volts, the three expressions become

$$v = 5\cdot947\ .\ 10^7\sqrt{V}(1 - 1\cdot474\ .\ 10^{-6}V),$$

$$\lambda = 1\cdot227\ .\ 10^{-7}\frac{1}{\sqrt{V}}(1 - 4\cdot914\ .\ 10^{-7}V),$$

and $\quad \lambda' = \dfrac{Dd}{37\cdot2}.$

If the wavelengths λ and λ' are equal equations (94) and (95) give the expression

$$d = \frac{2hl}{D\sqrt{2em_0V}}\left(1 - \frac{eV}{4m_0c^2}\right) \qquad . \qquad . \qquad (96)$$

for the spacing of the reflecting planes. For different potentials and for rings of the same indices, therefore, the quantity

$$D\sqrt{V}\left(1 + \frac{eV}{4m_0c^2}\right)$$

is in this case constant. We will denote the factor in brackets in this expression by r, i.e. $r = (1 + 4\cdot914\ .\ 10^{-7}V)$. The test of the equality of λ and λ' is the constancy of $rD\sqrt{V}$ (at a value equal to that of $10^{-4}\ .\ 2hl/d\sqrt{2em_0}$) in experiments at different potentials.

In taking the photographs required for comparing λ and λ' the method of the simultaneous spark alone was used, this being the most satisfactory way of using a spark gap for determining the potential V corresponding to a given diffraction ring. In each case the diameter used for the comparison was that of the innermost ring (D_1). In the case of gold D_1 is the diameter of the smallest ring which shows the four maxima, i.e. the ring formed by the (200) and (020) planes. It is the mean of the internal and external diameters measured near the horns of the maxima, where the radial width of the ring is small.

EXPERIMENTAL RESULTS

Gold. The results of measurements by this method for a gold film are exhibited graphically in Fig. 93 (Curve *A*), the horizontal axis representing the peak voltage applied to the electrodes of the discharge tube as indicated by the spark length, the vertical axis the values of $rD_1\sqrt{V}$ which, as we have seen, should be constant if the wavelengths λ and λ' are equal. The points marked ○ in Curve *A* represent the mean results

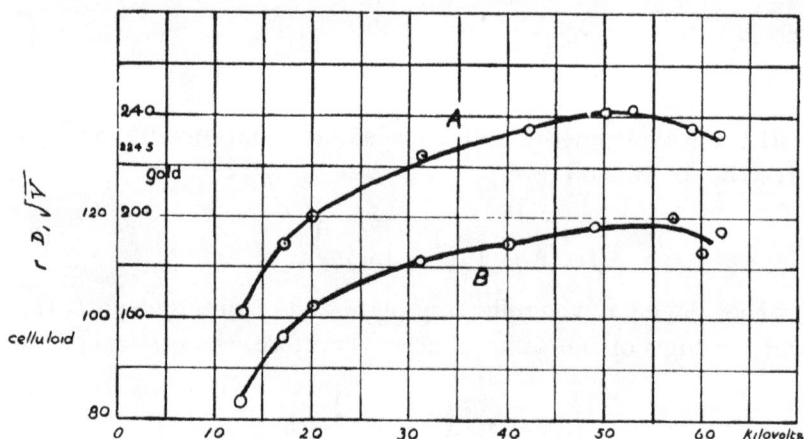

FIG. 93. VARIATION OF $rD_1\sqrt{V}$ WITH VOLTAGE
A — Gold.   *B* — Celluloid.

obtained from two sets of experiments, made with transmission through different parts of the same film, the two series of results not differing from each other at any voltage by more than 2·5 per cent in the value of $rD_1\sqrt{V}$.

From X-ray measurements it is known that the spacing d of the (200) planes in gold (i.e. one-half the side of the cubic cell) is 2·032 Å.U., so that the value of $2hl \cdot 10^{-4}/d\sqrt{2em_0}$ is 224·5. If the wavelengths λ and λ' are equal, Curve *A* in Fig. 93 should coincide with the horizontal straight line at this height above the horizontal axis.

So far from this being the case, Curve *A* shows a regular increase in the value of $rD_1\sqrt{V}$ from 162 at 12,600 volts to a maximum of 243 at about 53,000 volts from which it falls slightly at higher potentials. According to these experimental results $rD\sqrt{V}$ has the value 224·5 only at a voltage of about 31,000.

The results of the same experiments are shown in another way in Fig. 94, in which the two upper curves indicate the values of the wavelengths λ (calculated from the voltage by equation 94), and λ' (from the diameter of the first ring by equation 95). It will be seen that the wavelength λ diminishes

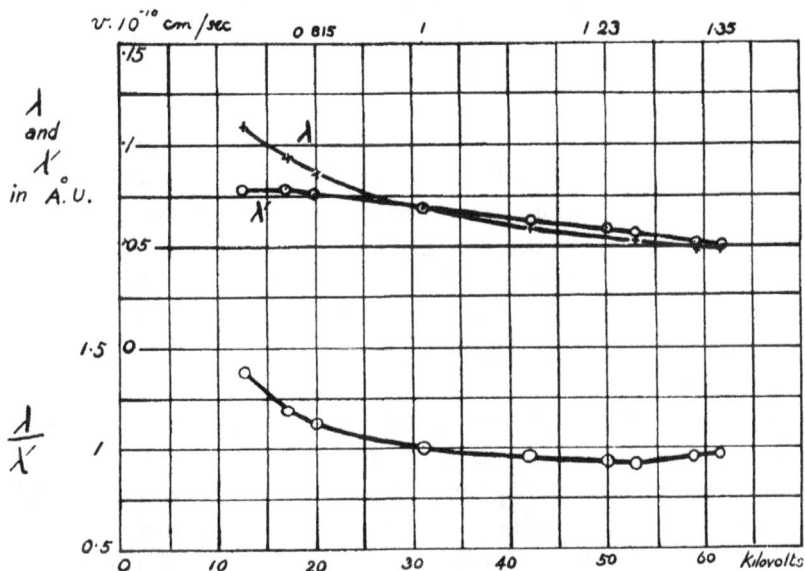

Fig. 94. Wavelengths Calculated from the Voltage and from the Gold Diffraction Patterns, and their Ratio

from 0·1085 Å.U. at 12,600 volts to 0.049 Å.U. at 60,000 volts, while λ' shows a much smaller variation over the same range of voltage, viz. from 0·0785 Å.U. to 0·05 Å.U. The lower curve in Fig. 94 represents the ratio of the two wavelengths, λ/λ', which diminishes from 1·38 at 12,600 volts to unity at 31,000, and further to a minimum of 0·92 at about 53,000, increasing slightly at higher voltages. Some values of the electron velocity v, calculated by equation (93), are indicated above the diagram of Fig. 94.

The large excess of λ over λ' at the lower potentials shown in Fig. 94, corresponding to the deficiency of $rD_1\sqrt{V}$ below the value 224·5 in Fig. 93, suggests first an inquiry into the degree of accuracy attained in the present experiments. With regard to the ring diameters, these could usually be measured to 0·1 mm., and there is no possibility of serious error in their measurement. As to the potential V the matter is not so simple. It has already been stated that in the present experiments sparks were passing, the gap having been previously adjusted to the required setting, while the photographs were being taken; but the question arises whether the potential does not frequently shoot up to a value much greater than the normal sparking value before the spark appears.

This effect occurs frequently in the discharge through gases at low pressure (see, for example, Figs. 76, 77, pp. 136, 137), and if we attempt to determine the maximum potential in such discharge by opening out a parallel gap until the spark just fails to appear, we shall probably find a value much greater than the potential at which the discharge through the tube normally passes. The maximum potential at the gap electrodes may, in fact, be much the same as if the tube were disconnected from them.

But this effect, known as "delayed discharge," is of very rare occurrence in the spark discharge between spherical electrodes in air at atmospheric pressure, and it may be safely assumed that when the spark passes simultaneously with the tube discharge the potential does not rise above the normal value corresponding to the gap setting. The potential may fail to rise to the sparking value (owing to variations in the degree of vacuum or to irregularity in the working of the interrupter), but in this event the diffraction rings are too large. In one experiment with a gold film, for example, the exposure was seventy-three flashes, at three of which the spark failed to appear. The photograph showed, in addition to the usual set of strong rings, a much weaker set of rather larger diameter doubtless produced by the three sparkless discharges.

The matter was also tested by the following experiment. It will be seen in the curve for λ' in Fig. 94 that the values of this wavelength, and, therefore, also the ring diameters, at the two lowest potentials applied in the experiments, are nearly equal, while the values of these potentials differ greatly, viz. 12,600 and 20,000 volts. These voltages correspond to the spark gap settings in the two experiments, viz. 3·5 and 6 mm. respectively. The experiment consisted in comparing the peak potentials with these sparks passing, the discharge-tube still connected to the spark gap electrodes, by the valve and electro-meter method.* The result was in close agreement with the ratio of the potentials indicated in Fig. 94, and since the ring diameters at these two potentials differ by only about 2 per cent, it must be concluded that within this range the diameter is not even approximately in the inverse ratio of the square root of the voltage applied to the tube when the photograph is taken.

It is remarkable that the change from 12,600 to 20,000 volts should have so little effect on the dimensions of the diffraction pattern. It is no greater than the difference of the ring diame-ters for different parts of the film and the *same* spark length, and, in fact, the variation of λ' with voltage over the whole range of the present experiments is comparatively small, being not more than one-half the variation of the other wavelength λ.†

As to the absolute values of the potential applied to the tube, these were obtained from a curve of sparking potentials, viz. the early part of the curve of Fig. 5, page 23. A spark gap is not regarded as an instrument of high precision for the measurement of potentials, but it is probably the most suitable for determining peak potentials in single discharges. In the writer's experience, a spark gap having spherical electrodes of zinc, occasionally cleaned by rubbing with fine emery cloth, is very suitable for determining variations of peak potential on a given occasion (see, for example, the curves of Figs. 14 to 19, pages 47–53), but its indications vary by about 2 or 3 per cent from day to day, probably owing to variations in atmospheric conditions. Such occasional variations are, how-ever, much too small to account for the large deviations of

* The form of electrometer used in this experiment consisted of two parallel vertical plates, one connected to the earthed side of the gap, the other, through a diode valve, to the negative electrode. A vertical wire connected with one of the plates was suspended about midway between them. The deflection of the wire measured by a microscope is, if small, proportional to the square of the potential difference of the plates, which, owing to the action of the valve, is the peak potential of the gap and discharge tube.

† M. Ponte (*Ann. de Physique*, 13, p. 395, 1930), by reflexion of electrons from zinc oxide crystals, has found very close agreement between λ and λ' over the range 7,000 to 20,000 volts.

$rD_1\sqrt{V}$ from constancy observed in the present experiments. An error of 38 per cent in $rD_1\sqrt{V}$, for example, would mean an error of 90 per cent in V.

Another point that should be considered in connection with the question of the accuracy of the measurements is whether the potential difference of the gap electrodes is, during the rise to the sparking value, the same as that of the terminals of the discharge tube. If this rise of potential took place with very great rapidity, there might be, owing to high frequency effects, an appreciable fall of potential in the wires connecting the gap to the tube, but it must be remembered that the rise of V_2 is comparatively slow, viz. that determined by the oscillation frequencies of the coil system. In these circumstances, there does not appear to be any reason for a considerable difference, up to the moment at which the spark appears, between the voltage of the gap and that of the discharge tube.

The conclusion to be drawn from the present experiments on gold films is, therefore, that while the two wavelengths λ and λ' are of the same order of magnitude, there is a regular variation in their ratio over the range of voltage 12,000 to 60,000, that is, over the range of electron velocity from $0\cdot65 \cdot 10^{10}$ cm./sec. to about $1\cdot35 \cdot 10^{10}$ cm./sec. Electrons arriving at the film with a velocity less than 10^{10} cm./sec. have their wavelength diminished in the film. This diminution might be accounted for on the very natural hypothesis that the electrons enter the regions of high positive potential within the atoms at which scattering takes place, and so have their momenta considerably increased. On this view the directions of the interference maxima in the scattered beam are determined by the wavelength when in this diminished state, i.e. the wavelength λ'.

On the other hand, the present results indicate that electrons having velocities between 10^{10} and $1\cdot35 \cdot 10^{10}$ cm./sec. have their wavelength increased in the film, as if the scattering took place in regions of *negative* potential. There appears to be no good reason for supposing that these more rapidly moving rays are scattered by electrons while those travelling more slowly are scattered by atomic nuclei, and the result, therefore, suggests that the electron wavelength is not determined entirely by the momentum as indicated by equation (90). It is difficult to say what other factors may influence it, but an extension of the experiments, especially to higher voltages, may throw some light upon the question.

It is clear that the variation of the ratio λ/λ' with voltage observed in the present experiments cannot be accounted for by the assumption that the electrons on entering the film simply come into a region of constant potential.

Celluloid. In measuring the rings obtained with celluloid films it was in some cases found that the distance between the centres of diametrically opposite spots in a ring was not quite definite, most of the spots lying on two rings of slightly different diameter. This may be due to the spacing of the reflecting planes differing slightly in adjacent parts of the film. Taking in each case the smaller diameter, the values of $rD_1\sqrt{V}$ at different voltages are shown in Fig. 93 (Curve B). This curve also shows a variation in the value of $rD_1\sqrt{V}$ similar to, though not quite so large or so regular as, that observed in gold. There is a large increase in the early part of the curve with some evidence of a diminution at the highest potentials.

If we assume that, as in gold, the wavelengths λ and λ' are equal at 31,000 volts, we find from equation (96) for d, the spacing of the planes in celluloid which form the first ring, the value 4·11 Å.U.

THE RATIOS OF THE DIAMETERS OF THE DIFFRACTION RINGS

Gold. The following are the ratios D_2/D_1, D_3/D_1, etc., of the diameters of the successive rings in the gold patterns to that of the first ring, with the probable correct values the first ring being taken as (200), and the plane indices to which they correspond. In the case of the larger rings allowance is made for the difference between $\tan 2\theta$ and $2 \sin \theta$—

$D_2/D_1 = 1 \cdot 411$ (mean of 15 plates, 12 giving values between
   1·4 and 1·43). Probably $\sqrt{8}/2$, (220).
        The second ring has four maxima on diameters at
        45° to those of the first ring. (See Fig. 90, page 159.)

D_3/D_1 = 1·67 (mean of 5). Probably $\sqrt{11}/2$, (311).

The third ring has eight rather faint maxima just outside the horns of those of the second ring.

D_4/D_1 = 2·00 (mean of 4). Probably the second order of (200), having its maxima on the same diameters. This ring is also faint.

D_5/D_1 = 2·18 (mean of 11). Probably $\sqrt{19}/2$, (331); maxima on the same diameters as those of (220).

D_6/D_1 = 2·88 (mean of 4). Probably $\sqrt{8}$, the second order of (220); maxima on the same diameters.

D_7/D_1 = 3·07 (mean of 2). Probably 3, the third order of (200); maxima on the same diameters.

D_8/D_1 = 3·58 (mean of 2). Probably $\sqrt{51}/2$, (551); maxima on same diameters as (220).

D_9/D_1 = 4·22 (mean of 2). Probably $3\sqrt{2}$, the third order of (220); maxima on same diameters.

In addition there is the very faint ring within the first (referred to on page 159), the diameter being $0·85D_1$ (approximately $\sqrt{3}D_1/2$), and another of three times this diameter. These two rings do not show maxima on the same diameters as (220), and therefore if they are due to (111) planes these planes do not belong to the main lattice, but are in a part of the film where the cubic cell is turned into a different position.

The ratios of diameters just given, and the positions of the maxima on the rings, show that all the principal rings of the gold examined are formed by the same lattice, having two faces of its cubic cell parallel to the surface of the film.

Celluloid. The corresponding results for celluloid films are as follows—

D_2/D_1 = 1·712 (mean of 18 plates giving values ranging from 1·67 to 1·77). Probably $\sqrt{3}$. The six maxima of the second ring are on diameters inclined at 30° to those of the first ring.

D_3/D_1 = 1·998 (mean of 10). Probably 2, the second order of D_1; maxima on same diameters.

$D_4/D_1 = 3.45$ (mean of 5). Probably $2\sqrt{3}$, the second order of D_2. The fourth ring is uniform in intensity round the circumference.

In addition to these four rings obtained with celluloid films, there is a fifth ring which is not included in the list because it did not appear in a complete form in any of the plates. The writer has, however, no doubt about the existence of this fifth ring, as diametrically opposite pairs of the spots upon it appear in some of the photographs. It is a faint ring of twelve spots, the mean diameter (from three plates) being $2.591\ D_1$. It is, therefore, intermediate in diameter between the third and fourth rings of the above list.

It has already been indicated that a ring having six maxima uniformly spaced round the circumference must be due to reflexion from three equally inclined sets of planes. If Fig. 95 represents a section of a celluloid film by a plane parallel to its surfaces, the reflecting planes are indicated by the three sets of straight lines in the diagram, the planes being at right angles to the surfaces of the film. If a scattering centre were placed at every point of intersection of the lines, with similarly situated centres in planes above and below the section indicated, all the five rings of the celluloid diffraction pattern would be accounted for. In such a uniform distribution of scattering centres, however, the molecules lose their identity, and for this reason the grouping indicated by the rings and spots in Fig. 95 is suggested.

In Fig. 95 the scattering centres are arranged in groups of six, each group forming a regular hexagon, and all the hexagons being arranged in planes parallel to the surfaces of the film. Groups in different layers are indicated by circles drawn round them, the full-line circles being in one plane, the broken-line circles in a layer above, and the dotted circles in a layer below the first. In the layer above that of the broken line circles, the grouping would be the same as that of the dotted circles, and in the next layer above, the same as that of the full-line circles, and so on. The grouping is, therefore, repeated regularly after three layers. This system of groups would give with considerable accuracy the observed ratios of diameters of the diffraction rings and the positions of the maxima upon them.

The three equally inclined sets of planes, parallel to *ab*, *bc*,

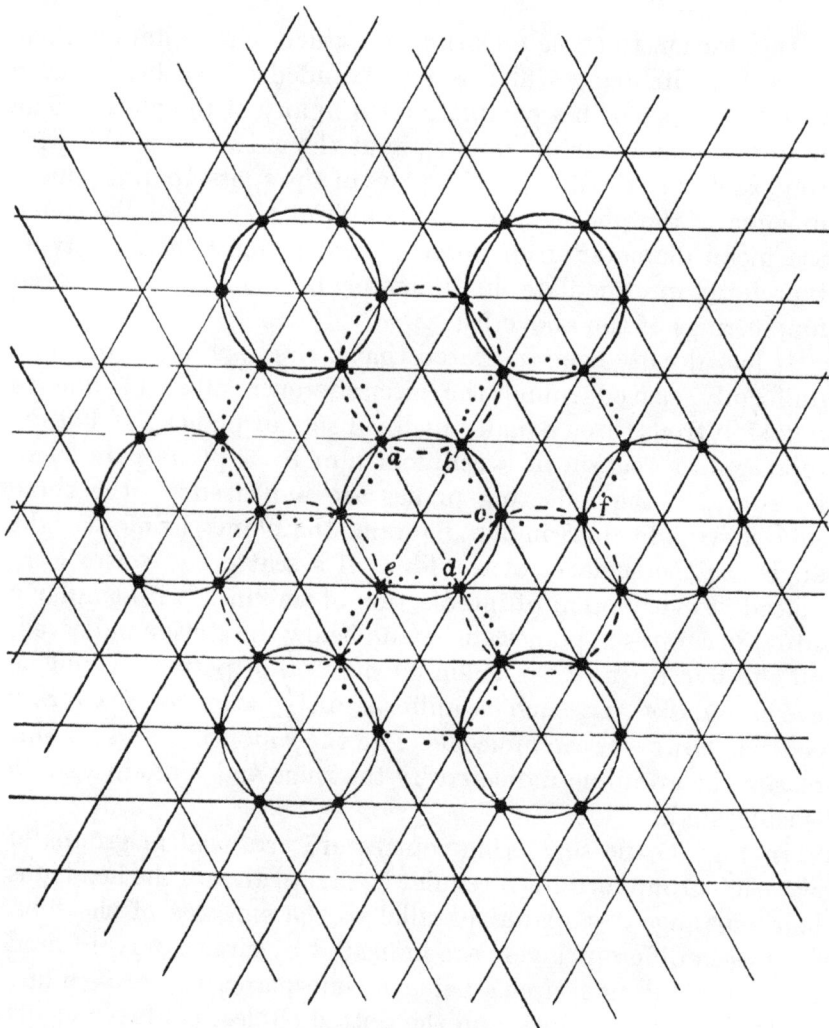

FIG. 95. SUGGESTED ARRANGEMENT OF MOLECULES IN A
CELLULOID FILM
The point *f* referred to on p. 173 is the first to the right of *c*.

and *cd* respectively are those which produce the six maxima of the first and third rings in the diffraction pattern. The spacing of this set of planes, in terms of *a* the side of a hexagon, is $a\sqrt{3}/2$, and if this be equated to 4·11 Å.U. (see page 169), we find $a = 4·75$ Å.U.

The planes of centres parallel to *ac, bd,* and *ce* form the second ring, with its maxima in the angles between those of the first ring, and by their second-order reflexion the fourth ring. The spacing of these planes is $a/2$, so that the diameter of the second ring should be $\sqrt{3}$ times that of the first ring, as observed. The planes of this second set are rather less rich in scattering centres than those of the first set, and, according to the grouping in Fig. 95, every third plane of the second set is vacant, with the consequence that the second ring in the pattern is considerably weaker than the first. In the photographs the maxima of the second ring are considerably weaker than those of the first, and this may be one reason for the difference of intensity, but there are also other factors which influence the relative intensity of the rings in a pattern.

The fifth ring in the celluloid pattern may be accounted for by reflexion from the planes parallel to *af* (Fig. 95) and similar sets. The spacing of these planes is $\frac{1}{2}a\sqrt{3/7}$, and the diameter of this ring should, therefore, be $\sqrt{7}$ times that of the first ring, i.e. $2·646\,D_1$. The ring should have twelve maxima, and it is probably represented by the incomplete ring observed in some of the photographs having a diameter of $2·591\,D_1$. One reason for the comparative faintness of the fifth ring is that the planes parallel to *af* have per unit area little more than one-third of the number of scattering centres in the planes which form the first ring.

It will be noticed that, according to the scheme of Fig. 95, all the five rings in the celluloid pattern can be accounted for by reflexions from planes which are at right angles to the surfaces of the film, and, therefore, very nearly parallel to the incident pencil. When the reflexions take place at such planes, there is another circumstance which tends to diminish the relative intensity of the larger rings in experiments of this kind. The larger rings require a greater angle of incidence upon the reflecting planes, and in the area covered by a very narrow incident normal pencil of cathode rays there will be comparatively few portions of the film in which these planes are inclined at a sufficiently large angle to the incident rays to produce strong reflexion. If the incidence were perfectly normal, there would be no reflexion from planes which are at right angles to the film.

The fact that the principal maxima in the celluloid rings number 6 or 12, but no more so far as these experiments show, suggests that the regularity of molecular arrangement is mainly confined to the portions of the film near its two surfaces. If this is the case, and if the chief reflecting planes near one surface are parallel to those near the other, there will be six principal maxima; if they are not parallel, there will be twelve maxima, uniformly spaced round the rings or arranged as six pairs, as shown in some of the photographs.

It has already been indicated that the arrangement of scattering centres suggested in Fig. 95 is not the only one which would produce diffraction rings showing the same ratios of diameters, and several others might be suggested. For example, the molecules might be regarded as linear chains of scattering centres having their lengths at right angles to the film, each chain being at a line of intersection of the planes shown in Fig. 95; or they might be arranged in columns of hexode groups all standing, for example, upon the full-line circles of Fig. 95. Such arrangements would give the observed ratios of diameters in the diffraction pattern, but it is difficult to understand why such chains or columns should be arranged at such regular distances from one another that the atoms (or groups of atoms) in them lie in definite planes throughout the substance, a condition which is necessary in order to account for the occurrence of diffraction rings with regularly spaced maxima round their circumferences. An arrangement such as that of Fig. 95, in which each hexode group, regarded as a unit, serves to hold together its three neighbours in an adjacent layer, seems to offer stronger dynamical reason for a uniformity of arrangement throughout the mass.

It is remarkable that such a substance as celluloid, which is usually regarded as amorphous, should give rise to diffraction rings having maxima, or spots, upon their circumferences as if the substance were crystalline, and so far as the writer is aware, no such evidence of crystalline structure has been found in this substance by means of X-rays. It has been suggested that the spots in the diffraction patterns produced by electron waves are due to the camphor which celluloid contains, but such "spot" photographs have been obtained with nitro-cellulose free from camphor by A. Dauvillier* and by F. Kirchner.† The same result has also been found by the present writer who, through the kindness of Dr. J. Weir, of Imperial Chemical Industries, was able to obtain some very pure specimens of nitro-cellulose and suitable solvents for this material. It was not found possible to prepare with this substance films quite so thin as those of celluloid, but films were obtained sufficiently thin to show rings with maxima upon them similar to those obtained with celluloid films, and of practically the same diameter for the same electron velocity. It must be concluded that the evidence of regularity of arrangement of the scattering centres is shown by nitro-cellulose in the form of thin films, in the complete absence of camphor or of other known crystalline substances. (See Appendix.)

With regard to the best wave-form of secondary potential in experiments of this kind, when an induction coil is used (with simultaneous spark) for obtaining single-flash photographs, the chief requirement is that the potential should rise quickly, and without secondary maxima, to the sparking value. It is desirable, therefore, that the frequency ratio of the coil should not be too great, and, therefore, also that the capacity of the condenser associated with the interrupter should be no greater than is necessary to ensure a good "break." The mercury dipper interrupter shown in Fig. 4, page 19, is probably the most suitable kind for the purposes of the experiment.

ADDITIONAL REFERENCES TO EXPERIMENTAL WORK ON
ELECTRON DIFFRACTION

C. J. Davisson and L. H. Germer: *Phys. Rev.*, 30, p. 705, 1927; *Proc. Nat. Acad. Sci.*, 14, pp. 317, 619, 1928.

G. P. Thomson: *Proc. Roy. Soc.*, A. 128, p. 649, 1930; A. 133, p. 1, 1931.

---

* *Comptes Rendus*, 191, p. 708 (1930).
† *Naturw.* 18, p. 706 (1930); 19, p. 463 (1931).

G. P. Thomson: *The Wave Mechanics of Free Electrons* (McGraw-Hill, 1930).

A. Reid: *Proc. Roy. Soc.*, A. 119, p. 663, 1928.

R. Ironside: *Proc. Roy. Soc.*, A. 119, p. 668, 1928.

E. Rupp: *Ann. d. Phys.*, 86, p. 981, 1928; 1, pp. 773, 801, 1929; 3, p. 497, 1929; *ZS. f. Phys.*, 58, p. 766, 1929; 61, pp. 158, 587, 1930.

S. Kikuchi: *Proc. Imp. Acad. Jap.*, 4, pp. 271, 275, 354, 471, 1928; *Jap. Journ. Phys.*, 5, p. 83, 1928; *Phys. ZS.*, 31, p. 777, 1930.

D. C. Rose: *Phil. Mag.*, 6, p. 712, 1928.

H. E. Farnsworth: *Phys. Rev.*, 34, p. 679, 1929; 37, p. 1,017, 1931.

F. Kirchner: *Phys. ZS.*, 31, pp. 772, 1,025, 1930.

R. Wierl: *Phys. ZS.* 31, pp. 366, 1,028, 1930; *Ann. d. Phys.*, 8, p. 521, 1931.

H. Mark and R. Wierl: *ZS. f. Phys.*, 60, p. 741, 1930; *Die experimentellen und theoretischen Grundlagen der Elektronenbeugung* (Borntraeger, Berlin, 1931).

CHAPTER VII

OTHER FORMS OF OSCILLATION TRANSFORMER

The Tesla Coil. The suggestion has been made by several experimenters that an induction coil might be worked by charging the primary condenser and discharging it through the coil, instead of by the interruption of a current in the primary circuit. The experiment appears to have been first tried by Norton and Lawrence,* who, with the aid of a rotary commutator, connected the condenser alternately to the 200-volt electric light mains and to the terminals of the primary coil.

A little consideration will show that this method of excitation is not likely to be attended with any great success in the case of an induction coil of the ordinary construction. Let us suppose, for example, that the capacity of the condenser is as great as 20 mfd. When charged to 200 volts its energy is 0·4 joule. If the self-inductance of the primary coil is 0·2 henry, the magnetic energy at a current of 10 amp. is 10 joules. Thus, in order to supply the coil with as much energy by charging the condenser, as can be easily supplied in the form of magnetic energy of the primary current, we should require either a very large condenser capacity, or an unusually high supply voltage. A very large primary capacity would, however, throw the system far out of adjustment—the capacity would be much greater than the optimum—unless the primary self-inductance were extremely small or the secondary terminals were connected with a large condenser. It is clear that, in general, no advantage is to be gained by supplying an induction coil with condenser charges rather than with an interrupted primary current.

On the other hand, in the arrangement of high-frequency coupled circuits known as the Tesla coil, the former method of excitation is the more suitable. In this case the primary coil consists of a few turns of wire—sometimes only a single turn—and has no iron core; consequently, an exceedingly

* *Electrical World*, 6th March, 1897, p. 327.

strong current would have to be supplied to the coil to represent
as much initial energy as can easily be supplied in the electrostatic form by charging a condenser. The condenser usually
takes the form of a Leyden jar, which is charged to the required
sparking potential by an induction coil. One terminal of the
primary of the Tesla coil is connected with a spark gap electrode, the other with a plate of the Leyden jar (see Fig. 96).
The other spark electrode and the other plate are connected
together. The terminals of the induction coil are connected

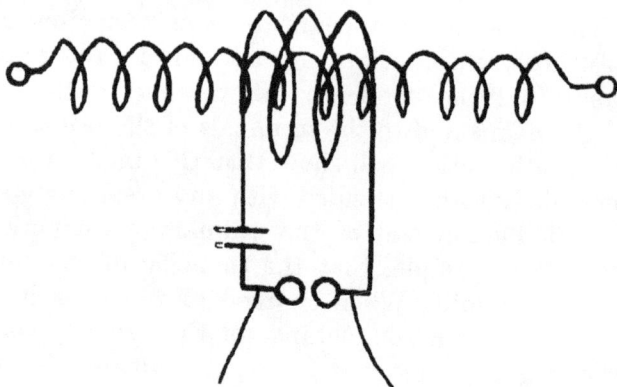

Fig. 96. Diagram of Circuits of Tesla Coil

with the plates of the jar or with the spark electrodes. Sometimes two jars are employed, these being connected in cascade
in the primary circuit of the Tesla coil. In this case the two
inner plates may be connected with the induction coil terminals
and with the spark electrodes, the two outer plates with the
terminals of the Tesla coil primary. In either case the primary
circuit of the Tesla coil includes the primary coil, the spark-
gap, and the jar (or the two jars in cascade). The secondary
coil usually consists of a much larger number of turns wound
in a single layer on an insulating tube or frame. The insulation
must be carefully attended to, otherwise sparks will pass
between the primary and secondary windings or between
neighbouring turns of the secondary; for this reason the coils
are sometimes immersed in oil.*

---

* Further information as to the construction of Tesla coils will be found
in Fleming's *Principles of Electric Wave Telegraphy and Telephony*, 2nd Ed.,
pp. 74–80.

We shall consider the adjustment of the Tesla coil system for maximum secondary potential. The fundamental equations are of the same form as (18) and (19), Chapter I, the coefficients here referring to the primary and secondary circuits of the Tesla coil, and V_1, V_2 being respectively the difference of potential of the plates of the jar and that of the terminals of the secondary coil. The initial conditions differ, however, from those which apply in the case of an induction coil worked by an interrupter. In the present problem they are: $V_1 = V_0$ (the sparking potential of the primary condenser), $V_2 = 0$, $i_1 = 0$, $i_2 = 0$, when $t = 0$. As in other instances of coupled oscillating circuits, the wave of potential in each circuit consists in general of two oscillations differing in frequency and damping factor, the frequencies and damping factors being, however, the same for both circuits.

If we neglect resistances, the solution for V_2 is—

$$V_2 = \frac{L_{21} C_1 V_0}{\sqrt{\{(L_1 C_1 - L_2 C_2)^2 + 4k^2 L_1 C_1 L_2 C_2\}}} \times (\cos 2\pi n_1 t - \cos 2\pi n_2 t), \qquad . \qquad . \qquad (97)$$

where n_1, n_2 are the frequencies of the system.

The primary potential difference is given by—

$$V_1 = V_0 \frac{n_1^2 n_2^2}{n_2^2 - n_1^2} \left[\frac{1 - 4\pi^2 n_2^2 L_1 C_1}{n_2^2} \cos 2\pi n_1 t - \frac{1 - 4\pi^2 n_1^2 L_1 C_1}{n_1^2} \cos 2\pi n_2 t \right]^* . \qquad . \qquad (98)$$

The full expression for V_2, in which the resistances are included, was given by Drude in a well-known paper,† but the above expressions will serve for the present purpose.

Drude considers the problem in which the secondary coil is given, and in which it is required to find what arrangement of the primary circuit gives the highest secondary potential.‡ After remarking (*l.c.*, p. 539) that it is necessary to distinguish between the two cases in which (*a*) the capacity C_1 is varied,

---

* These expressions may be deduced from the solutions given by Fleming, *l.c.*, p. 264.

† P. Drude, *Ann. d. Physik*, xiii, p. 512 (1904).

‡ The primary sparking potential V_0 is also supposed to be given.

and (*b*) the self-inductance L_1 is the variable quantity, he apparently comes to the conclusion (p. 540) that in either case the highest secondary potential is attained when L_1C_1 is equal (or nearly equal) to L_2C_2, i.e. when the periods of the primary and secondary circuits, separated from each other, are equal. This is the case of so-called "resonance," and all the subsequent calculations and conclusions given by Drude, including the tables and curves (*l.c.* pages 546–551) are based on the assumption that this condition ($L_1C_1 = L_2C_2$) is satisfied. Drude finally arrives at the result that, if the damping of the oscillations is small, the secondary potential is greatest when the coupling coefficient is 0·6 and the primary capacity is so adjusted as to bring the primary circuit into "resonance" with the secondary, this adjustment of the system giving the frequency-ratio $n_2/n_1 = 2$.

The reasoning by which Drude arrives at the above result is not perfectly clear, and the result does not hold in case (*a*). Denoting the ratio L_2C_2/L_1C_1 by m, the expression (97) for V_2 becomes—

$$V_2 = \frac{L_{21}V_0}{L_1\sqrt{(1-m)^2 + 4k^2m}}(\cos 2\pi n_1 t - \cos 2\pi n_2 t)* \qquad . \qquad (99)$$

If the primary capacity C_1 alone is varied (L_1, L_{21}, k^2, V_0, and L_2C_2 being constant), the denominator of this expression has a minimum value when—

$$m = 1 - 2k^2 \qquad . \qquad . \qquad . \qquad . \qquad . \qquad (100)$$

If also n_1 and n_2 are suitably related, the maxima of the two waves in the secondary coil will occur simultaneously, so that at time $1/2n_1$, $\cos 2\pi n_1 t = -1$, $\cos 2\pi n_2 t = 1$. This happens, for example, when $k^2 = 0·265$,† $m = 1 - 2k^2 = 0·47$, which adjustment makes the frequency-ratio n_2/n_1 equal to 2.

It follows from (99) that if $k^2 = 0·265$ the most effective primary capacity is that which makes $m = 0·47$, i.e. $L_1C_1 = 2·128\ L_2C_2$. At this degree of coupling, therefore, the optimum primary capacity should be more than twice as great

---

* *Phil. Mag.*, 30, p. 236 (1915).   † More exactly, $k^2 = 9/34$.

as the value necessary for "resonance." The maximum value of V_2 in this case is numerically, by (99)—

$$V_{2m} = 2 \cdot 267 \frac{L_{21} V_0}{L_1}{}^* .$$

The adjustment recommended by Drude ($k^2 = 0 \cdot 36$, $L_1 C_1 = L_2 C_2$) gives for the maximum secondary potential –

$$V_{2m} = 1 \cdot 667 \frac{L_{21} V_0}{L_1} .$$

Even at this degree of coupling, however, a higher secondary potential may be obtained with a larger primary capacity. For example, the adjustment $k^2 = 0 \cdot 36$, $L_1 C_1 = 2 L_2 C_2$, gives the frequency-ratio $n_2/n_1 = 2 \cdot 197$, and a maximum secondary potential, at time $t = 0 \cdot 925/2n_1$, the value of which is

$$V_{2m} = 1 \cdot 998 \frac{L_{21} V_0}{L_1} .$$

From these examples it will be seen that Drude's rule does not in general give even approximately the correct value of the optimum primary capacity. The most effective capacity is generally greater than the "resonance" value.

The matter has been examined experimentally by W. Morris Jones,[†] who determined the optimum primary capacity for a Tesla coil for various degrees of coupling ranging from $k^2 = 0 \cdot 115$ to $k^2 = 0 \cdot 374$. The measurements were made both with the primary coil over the middle of the secondary (secondary terminals insulated), and with the primary at the lower end of the secondary (lower secondary terminal earthed). The secondary potential was indicated by the distance, measured from a long wire connected with the upper terminal of the secondary coil, at which a neon tube just failed to glow. The coupling was varied by removing turns from the secondary coil, measurements being made (by wave-meter methods), at each stage of the coupling, and of the ratios $L_2 C_2/L_1 C_1$ and n_2/n_1, when the primary capacity was the optimum. The

---

[*] The adjustment $k^2 = 0 \cdot 265$, $m = 1$, gives approximately $V_{2m} = 1 \cdot 88 L_{21} V_0/L_1$.

[†] *Phil. Mag.*, p. 62, Jan., 1916.

results of these experiments showed that in all the cases examined the ratio $L_2 C_2 / L_1 C_1$ was considerably less than unity, i.e. that the optimum capacity was considerably greater than the "resonance" value. The ratio of the optimum to the resonance value was found to increase with the coupling, in one case being as great as 4. The frequency-ratio (for optimum adjustments) also increased with the coupling and passed through 2 at a value of the coupling k^2 not far from 0·265.

The experiments thus agree with the above theory in showing that the highest secondary potential is (except at very loose coupling) obtained with a primary capacity considerably greater than that required to make the periods of the two circuits equal when separated.

The conditions are very different if the primary capacity is kept constant and L_1 is varied, e.g. by means of variable series inductance in the primary circuit. In this case L_{21} is constant, k^2 is inversely proportional to L_1, and the coefficient of $(\cos 2\pi n_1 t - \cos 2\pi n_2 t)$ in (99) has a maximum value of

$$\frac{V_0}{2}\sqrt{\frac{L_{21}}{L_{12}} \cdot \frac{C_1}{C_2}}$$ when $m = 1$. If, in addition, $n_2 = 2n_1 (k^2 = 0\cdot36)$ the numerical maximum of V_2 is given by—

$$V_{2m} = V_0 \sqrt{\frac{L_{21}}{L_{12}} \cdot \frac{C_1}{C_2}},$$

consequently,

$$\tfrac{1}{2}C_2 V_{2m}{}^2 \cdot \frac{L_{12}}{L_{21}} = \tfrac{1}{2}C_1 V_0{}^2,$$

i.e. the maximum electrostatic energy in the secondary coil is equal to the initial energy in the primary condenser.

In Drude's adjustment, therefore, the efficiency is unity if the resistances are negligible. This result may also be seen from the expression (98) for the primary potential, which reduces, if $\cos 2\pi n_1 t = -1$, $\cos 2\pi n_2 t = 1$, to

$$V_1 = V_0 \frac{m-1}{\sqrt{(m-1)^2 + 4k^2 m}}$$

and, therefore, vanishes when $m = 1$. At the moment when the secondary potential reaches its maximum value, therefore, the primary condenser is uncharged, and, since $\dfrac{dV_1}{dt} = 0$, $\dfrac{dV_2}{dt} = 0$, there is no current in either circuit, so that the whole of the energy exists as electrostatic energy in the secondary coil.

In this kind of variation, in which L_1 alone is varied, the initial energy supplied to the system is constant, and the adjustment which gives maximum efficiency must also give the greatest secondary potential. But when the primary capacity is increased the energy supplied to the system becomes greater, and the secondary potential may be increased to a certain extent beyond the point corresponding to maximum efficiency.

Both forms of the problem may present themselves in practice. If, for example, the Tesla coil is given, and ample energy is available for charging the condenser, the capacity of the condenser may always be adjusted to the "optimum" value, whatever be the value of k^2. But if, on the other hand, the energy is limited and it is required to construct a Tesla coil which will make the fullest use of it in generating high secondary potential, the primary capacity is then constant, viz. the greatest that can be charged to the required sparking potential with the energy available at each discharge of the induction coil. In this case maximum efficiency and small secondary capacity should be the chief considerations borne in mind in the construction of the Tesla coil.

The Auto-Transformer. A very convenient arrangement for producing high-tension high-frequency electrical oscillations is that known as an auto-transformer, which consists of a single coil, a few of the turns of which act as primary and the rest of the coil as secondary. In one form* the coil consists of 150 to 200 turns of bare copper wire of about 1·5 mm. diameter, wound on an ebonite frame 3 to 4 ft. high and 9 to 12 in. in diameter. A few turns at the lower end of the coil are connected to the spark gap and condenser, as shown in Fig. 97, the position of the contact J being adjustable.

The theory of the auto-transformer differs slightly from that of the Tesla coil because the primary coil is connected to the

---

* Also known as an "Oudin resonator."

secondary, so that the coupling in this arrangement is both electrical and magnetic. If the difference of potential between the ends of the whole coil be denoted by V_2, that of the plates of the condenser by V_1, the current flowing into the secondary

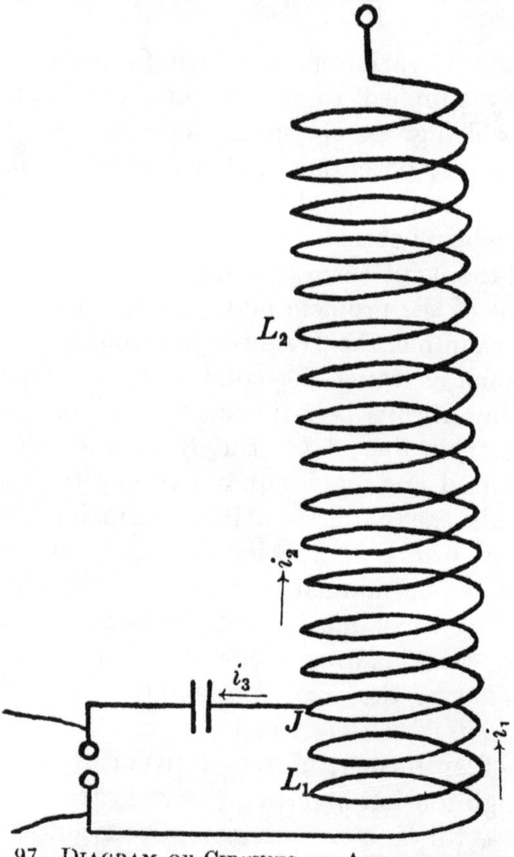

FIG. 97. DIAGRAM OF CIRCUITS OF AUTO-TRANSFORMER

at J by i_2, the condenser current by i_3, and with the usual notation for the other quantities as indicated in Fig. 97, the equations for the primary and secondary circuits (neglecting resistances) are—

$$L_1 \frac{di_1}{dt} + I_{12} \frac{di_2}{dt} + V_1 = 0 \quad . \quad . \quad . \quad . \quad . \quad (101)$$

$$L_2 \frac{di_2}{dt} + L_{21} \frac{di_1}{dt} + V_2 - V_1 = 0 \quad . \quad . \quad . \quad (102)$$

with the conditions

$$i_2 = C_2 \frac{dV_2}{dt} \qquad \qquad \text{. . . . . . . (103)}$$

$$i_3 = C_1 \frac{dV_1}{dt} \qquad \qquad \text{. . . . . . . (104)}$$

$$i_1 = i_2 + i_3$$
$$= C_1 \frac{dV_1}{dt} + C_2 \frac{dV_2}{dt}. \qquad \text{. . . . . (105)}$$

The currents being eliminated, equation (101) becomes—

$$L_1 C_1 \frac{d^2 V_1}{dt^2} + (L_1 + L_{12}) C_2 \frac{d^2 V_2}{dt^2} + V_1 = 0 \qquad . \qquad . \text{ (106)}$$

and this added to (102) gives—

$$(L_1 + L_{21}) C_1 \frac{d^2 V_1}{dt^2} + s L_2 C_2 \frac{d^2 V_2}{dt^2} + V_2 = 0 \qquad . \qquad . \text{ (107)}$$

where s is a fraction, slightly greater than unity, defined by—

$$s L_2 = L_1 + L_{12} + L_{21} + L_2 \qquad . \qquad . \qquad . \text{ (108)}$$

The assumed solutions $V_1 = A e^{\iota p t}$, $V_2 = B e^{\iota p t}$, substituted in (106) and (107), lead, after elimination of the ratio B/A, to the equation for $p(= 2\pi n)$

$$p^4 L_1 C_1 L_2 C_2 (1 - k^2) - p^2 (L_1 C_1 + s L_2 C_2) + 1 = 0 \qquad . \text{ (109)}$$

in which k^2 is the magnetic coupling $L_{12} L_{21} / L_1 L_2$. The two frequencies, n_1, n_2, of the system are, therefore, given by the equation—

$$8\pi^2 n^2 (1 - k^2) = \frac{1}{L_2 C_2} + \frac{s}{L_1 C_1}$$
$$\pm \sqrt{\left(\frac{1}{L_2 C_2} + \frac{s}{L_1 C_1} \right)^2 - \frac{4(1 - k^2)}{L_1 C_1 L_2 C_2}} \qquad . \qquad . \text{ (110)}$$

In the extreme case, $C_1 = \infty$ (primary closed), one of the frequencies is zero and the other is $1/2\pi \sqrt{L_2 C_2 (1 - k^2)}$. In the other extreme case, $C_1 = 0$, one frequency is infinite and the other is $1/2\pi \sqrt{s L_2 C_2}$, that is, it is the frequency of the primary and secondary oscillating together as one coil. The ratio of the squares of the frequencies in these two cases is $(1 - k^2)/s$, a result which suggests an experimental method for determining the coupling. In general, if we write u for the ratio L_1C_1/L_2C_2, the frequency ratio is given by—

$$\frac{n_2^2}{n_1^2} = \frac{s + u + \sqrt{(s+u)^2 - 4(1-k^2)u}}{s + u - \sqrt{(s+u)^2 - 4(1-k^2)u}} \qquad . \qquad . \qquad . \quad (111)$$

For any given value of k^2 the frequency ratio is smallest when $u = s$, i.e. when $L_1C_1 = sL_1C_1$.

In order to find the amplitudes multiply (107) by any factor λ and add its terms to those of (106). We then have the equation—

$$\left\{L_1 + \lambda(L_1 + L_{21})\right\} C_1 \frac{d^2V_1}{dt^2}$$
$$+ \left\{L_1 + L_{12} + \lambda sL_2\right\} C_2 \frac{d^2V_2}{dt^2} + V_1 + \lambda V_2 = 0.$$

If λ is so chosen that—

$$(L_1 + L_{12} + \lambda sL_2)C_2 = \lambda\left\{L_1 + \lambda(L_1 + L_{21})\right\} C_1 \qquad . \quad (112)$$

then

$$\left\{L_1 + \lambda(L_1 + L_{21})\right\} C_1 \frac{d^2}{dt^2}(V_1 + \lambda V_2)$$
$$+ (V_1 + \lambda V_2) = 0. \qquad . \qquad . \qquad . \qquad . \quad (113)$$

The two values of λ, viz. λ_1 and λ_2, may be calculated by (112) in terms of the coefficients of equations (106) and (107). They may also be expressed in terms of the frequencies n_1 and n_2, for, by (113)—

$$\frac{1}{4\pi^2n_1^2} = \left\{L_1 + \lambda_1(L_1 + L_{21})\right\} C_1$$
$$\frac{1}{4\pi^2n_2^2} = \left\{L_1 + \lambda_2(L_1 + L_{21})\right\} C_1.$$

Thus,

$$\left.\begin{aligned}
\lambda_1 &= \frac{1}{(L_1 + L_{21})C_1}\left(\frac{1}{4\pi^2n_1^2} - L_1C_1\right) \qquad . \qquad . \\
\lambda_2 &= \frac{1}{(L_1 + L_{21})C_1}\left(\frac{1}{4\pi^2n_2^2} - L_1C_1\right) \qquad . \qquad .
\end{aligned}\right\} \quad . \quad (114)$$

The solution of equation (113) is represented by the two normal vibrations—

$$V_1 + \lambda_1 V_2 = A_1 \sin (2\pi n_1 t + \delta_1)$$
$$V_1 + \lambda_2 V_2 = A_2 \sin (2\pi n_2 t + \delta_2) \qquad \Big\} \quad . \ (115)$$

The solutions for V_2 and V_1 are, therefore—

$$V_2 = \frac{A_1}{\lambda_1 - \lambda_2} \sin (2\pi n_1 t + \delta_1) - \frac{A_2}{\lambda_1 - \lambda_2} \sin (2\pi n_2 t + \delta_2) \ (116)$$

$$V_1 = \frac{A_1 \lambda_2}{\lambda_2 - \lambda_1} \sin (2\pi n_1 t + \delta_1) - \frac{A_2 \lambda_1}{\lambda_2 - \lambda_1} \sin (2\pi n_2 t + \delta_2) \ (117)$$

The coefficients A_1, A_2, and the phase angles δ_1, δ_2, are to be determined from the initial conditions. These express that at the moment ($t = 0$) at which the spark appears—

$V_1 = V_0$, the sparking potential,

$V_2 = 0$,

$i_1 = 0$, $\quad i_2 = 0$,

and, therefore,

$$\frac{dV_1}{dt} = 0, \quad \frac{dV_2}{dt} = 0.$$

On substituting these values in (116) and (117), we find—

$$A_1 = A_2 = V_0,$$

$$\delta_1 = \delta_2 = \frac{\pi}{2},$$

and the solutions for V_2 and V_1 are, therefore—

$$V_2 = \frac{V_0}{\lambda_1 - \lambda_2} \Big(\cos 2\pi n_1 t - \cos 2\pi n_2 t \Big)$$

$$V_1 = \frac{V_0}{\lambda_2 - \lambda_1} \Big(\lambda_2 \cos 2\pi n_1 t - \lambda_1 \cos 2\pi n_2 t \Big)$$

Inserting the values of λ_1, λ_2 from (114), the expressions become—

$$V_2 = \frac{4\pi^2 (L_1 + L_{21}) C_1 V_0 n_1^2 n_2^2}{n_2^2 - n_1^2} \Big(\cos 2\pi n_1 t - \cos 2\pi n_2 t \Big)$$

$$V_1 = V_0 \cdot \frac{n_1{}^2 n_2{}^2}{n_2{}^2 - n_1{}^2} \left[\frac{1 - 4\pi^2 n_2{}^2 L_1 C_1}{n_2{}^2} \cos 2\pi n_1 t \right.$$
$$\left. - \frac{1 - 4\pi^2 n_1{}^2 L_1 C_1}{n_1{}^2} \cos 2\pi n_2 t \right].$$

In terms of the inductances and capacities of the system, the expression for the secondary potential V_2 is, by (110)—

$$V_2 = \frac{V_0 (L_1 + L_{21}) C_1}{\sqrt{\{(L_1 C_1 + s L_2 C_2)^2 - 4(1 - k^2) L_1 C_1 L_2 C_2\}}}$$
$$(\cos 2\pi n_1 t - \cos 2\pi n_2 t)$$
$$= \frac{(L_1 + L_{21}) V_0}{L_1 \sqrt{\{(1 + sm)^2 - 4(1 - k^2)m\}}}$$
$$(\cos 2\pi n_1 t - \cos 2\pi n_2 t) \qquad . \qquad . \ (118)$$

if m be written for the ratio $L_2 C_2 / L_1 C_1$.

As in the Tesla coil, the two oscillations of secondary potential in an auto-transformer have equal amplitudes and they begin in opposite phases. The expression (99) of the previous section may, in fact, be deduced from (118) by setting s equal to unity. The variation of V_2 with the adjustment of the system follows the same course in the auto-transformer as in the Tesla coil, with small differences in the numerical values depending upon the fraction s. For example, the amplitude of the V_2 oscillations is a maximum (all the inductances being given) when—

$$m = \frac{2(1 - k^2) - s}{s^2},$$

and this condition is satisfied, with n_2/n_1 equal to 2, at the coupling—

$$k^2 = 1 - s \frac{25}{34}.$$

In general, for a given position of the contact J (Fig. 97), the most effective primary capacity is considerably greater than the value required to make $L_1 C_1 = L_2 C_2$.

If the resistance terms had been retained in the equations for the circuits of a Tesla coil or an auto-transformer, the expressions for the oscillations in each circuit would have contained factors $e^{-k_1 t}$ and $e^{-k_2 t}$ representing the decay of the amplitudes. The expressions for the damping factors k_1 and k_2 are identical with those given in equations (46), (47), and (48) on page 85, with the exception that in the case of the auto-transformer L_2 in equation (48) must be replaced by sL_2.

The High Tension Magneto. The high tension magneto was a few years ago much used for ignition in motor-car and aeroplane engines, but it has now been largely superseded in cars by the coil and battery system, which is found to be more convenient for use in connection with multi-cylinder engines.

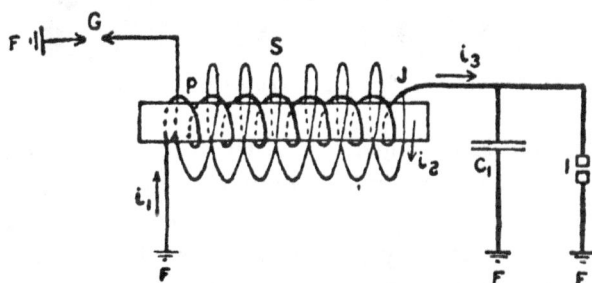

Fig. 98. Circuits of Magneto

The arrangement of the circuits of a magneto is shown diagrammatically in Fig. 98, in which P is the primary coil, S the secondary, both being wound on a laminated iron core. I is the contact breaker, C_1 the primary condenser, and G the sparking plug. The points F are connected to the frame of the machine.

In one type of magneto, the core upon which the coils are wound, and to which also the condenser C_1 and the contact breaker are attached, forms an armature which is rotated between the poles of a permanent magnet so that alternations of magnetic flux are produced in the core which serve to generate the primary current during the periods when there is contact at I. The photograph reproduced in Fig. 99, taken with the current oscillograph connected in the primary circuit, shows the manner in which the primary current varies between a "make" and the following "break." After a very gradual beginning, the current rises rapidly to its maximum value, from which it falls slowly to the value at which "break" occurs. In this experiment the armature was rotating at about 1,800 revolutions per minute, the maximum current indicated being about 3 amp., and the contact breaker being arranged to make one interruption per revolution. In actual use, the

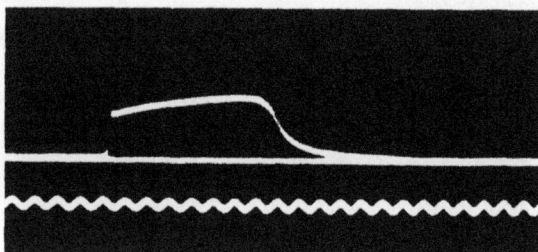

FIG. 99. OSCILLOGRAM OF PRIMARY CURRENT OF MAGNETO

contact breaker is arranged so as to interrupt the current at an earlier stage, viz. near the point at which the current has its maximum value. Fig. 100 shows in section the position of the armature very shortly before the moment at which "break" occurs. Owing to the form of the armature core and the variation during the rotation of its position with reference to the pole pieces, the inductances of the primary and secondary coils

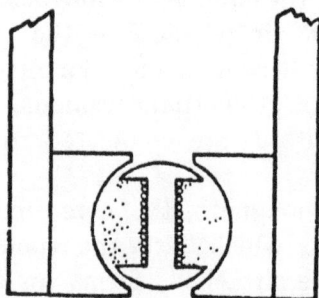

FIG. 100. ARMATURE IN POSITION OF MAXIMUM INDUCTANCES

are different in different positions of the armature. The inductances have their greatest values when the armature is in the position shown in Fig. 100.

It will be seen from Fig. 98 that the primary and secondary circuits of a magneto are connected at the point *J*, and it follows that, the currents being represented as shown in Fig. 98, the equations of the circuits are the same as those of the auto-transformer (see Fig. 97), with the addition of terms representing the electromotive forces induced in the primary and secondary coils owing to their rotation in the field of the permanent magnet. The E.M.F. due to the rotation is, however, usually small in comparison with the potential produced in the secondary coil at the interruption of the primary current.

When resistances are neglected, the circuits of a magneto oscillate with two frequencies given by equation (110). The theory of the secondary potential produced at "break" is very similar to that described in Chapter II for the case of the induction coil, and it can be shown that, corresponding to equation (42), page 37, the maximum secondary potential of a magneto is given by the expression—

$$V_{2m} = \frac{(L_1 + L_{21})\, i_0}{\sqrt{L_2 C_2}}\, U \sin \varphi, \qquad . \qquad . \qquad . \qquad . \quad (119)$$

where $\quad U = \dfrac{1}{\sqrt{u + s - 2\sqrt{(1 - k^2)u}}},$

and the angle φ is given by equation (41), page 36. For a certain series of values of the coupling k^2 the angle φ, determining the phase at which the maximum potential occurs, is $\pi/2$, and in these cases the maximum is given by—

$$V_{2m} = i_0 \sqrt{\frac{L_1}{C_2}} \sqrt{\frac{L_1 + L_{21}}{L_1 + L_{12}}},$$

which expression (with the potential due to the rotation added to it) represents the maximum theoretical potential attainable in any magneto. It follows also from (119) that in a magneto, as in an induction coil, the relation between the maximum secondary potential and the capacity of the primary condenser (or between $U \sin \varphi$ and u) is represented by a curve consisting of a series of arches similar to those described in Chapter II, the precise form of the curve depending upon the coupling of the primary and secondary circuits.

In a magneto, however, the damping of the oscillations, especially that of the more rapid component, is very great, owing chiefly to eddy current losses in the iron core and leakage from the secondary winding, and it is doubtful whether, without some modification of the circuits, the more rapid component ever exists as an oscillation, or whether it is not, owing to the excessive damping, replaced by aperiodic components.

It appears, however, that a very slight modification of the circuits is sufficient to bring the two oscillations well into evidence, as shown by the arches of the capacity-potential curves, or in other ways. Fig. 101, for example, shows a curve

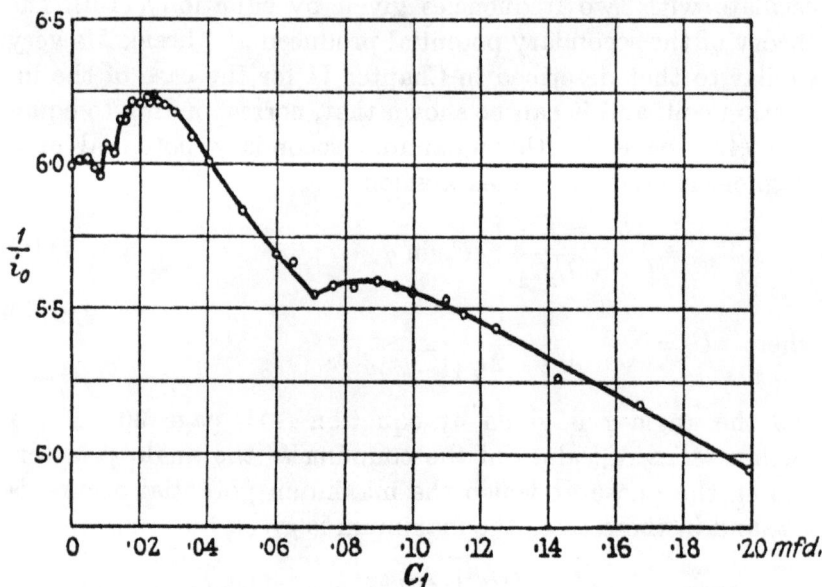

FIG. 101. VARIATION OF SECONDARY VOLTAGE WITH PRIMARY CAPACITY IN MAGNETO

indicating the variation of maximum secondary potential (observed by means of a spark gap) with primary capacity, the only modification of the circuits in this experiment being that a coil of very small self-inductance (about 0·0002 henry) was connected in series with the primary coil of the magneto. The curve has well-marked arches, proving the existence of both oscillations, and the relative proportions of the two principal arches indicate a coupling of about 0·78.

In Fig. 102 is shown a wave-form of secondary potential of a magneto. The photograph was obtained with the electrostatic oscillograph, modified by reduction of the length of the strip to 3 mm. The curve shows both oscillations, the frequency of the more rapid component being about 10,000 per sec., and about seven times that of the slower component. In this experiment, however, there was a very small series inductance (about 0·00002 henry) in the primary circuit, and a condenser in the secondary. Without such additions to the circuits, the magneto wave-form shows only aperiodic variations.

Fig. 103 is a reproduced photograph of a magneto spark, taken with a rotating mirror, and showing very clearly the oscillation of the system as modified by the secondary discharge. The discharge is of the "pulsating arc" kind (see also

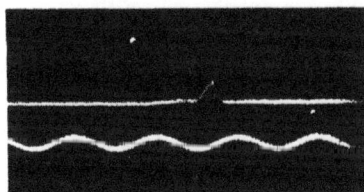

FIG. 102. DOUBLE OSCILLATION OF MAGNETO CIRCUITS

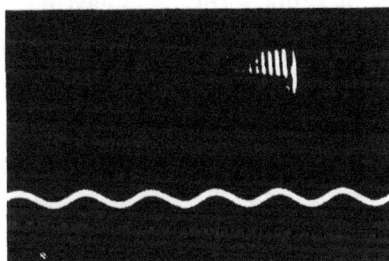

FIG. 103. MAGNETO SPARK

Fig. 61), the interval between two consecutive bands in the discharge being equal to $2\pi\sqrt{L_1C_1(1 - k^2)}$. In the example illustrated in Fig. 103, a 1-mfd. condenser was connected in parallel with the magneto condenser, the total primary capacity being 1·208 mfd., and the frequency of the pulsations is 6,645. When the 1-mfd. condenser was removed, the frequency of the pulsations was found from the photograph to be 16,010. The ratio of the frequencies is thus 2·409, which is also the square root of the inverse ratio of the capacities, i.e. $\sqrt{1·208/0·208}$.

In some respects, therefore, and with slight modifications in its circuits, the magneto behaves closely in accordance with the general theory of oscillation transformers. Regarded as a machine for producing electrostatic energy in the secondary circuit, the magneto is, however, probably the least efficient of all types of oscillation transformer. Probably not more than about 36 per cent of the magnetic energy $\frac{1}{2}L_1i_0^2$ in the primary circuit at break is converted into secondary electrostatic energy, and, according to the measurements made by Mr. Elwyn Jones,* only about 50 per cent of the work done in rotating the armature appears as primary magnetic energy. The total efficiency of the machine is about 18 per cent. The unavoidably low efficiency of conversion is one of the principal defects of the magneto, the rather strong primary currents required in the working of the machine being apt to cause deterioration of the interrupter contact surfaces. In spite of this, and owing to the skill and care displayed by the manufacturers in its construction, the magneto has proved itself to be a remarkably serviceable instrument.

* *Phil. Mag.*, p. 386, September (1923).

CHAPTER VIII

SPARK IGNITION*

ONE of the most important practical purposes for which the induction coil is used at the present time is that of the production of the spark required for ignition in internal combustion engines, and it is partly in this connection that attention has been drawn to the study of spark ignition, the main object of the inquiry being to discover what property of an electric spark determines whether it is or is not capable of causing general inflammation of an explosive mixture of gases, and what kind of discharge is the most effective in causing ignition.

Thermal Theory. With regard to the first of these questions, the oldest and, perhaps, the most natural hypothesis is that the igniting power of a spark depends simply upon the quantity of heat produced in it. This heat, the equivalent of the electrical energy dissipated in the spark, is, if the spark electrodes are thin rods or wires, mainly spent in raising the temperature of the surrounding gas, and it seems reasonable to suppose that if the gas, or a sufficient volume of it, is raised to the "ignition temperature," general inflammation will follow.

Ignition by Capacity and by Inductance Sparks. Experimental measurements of the heat developed in sparks just capable of producing ignition lead, however, to results which at first sight appear to be inconsistent with this view of the matter. It was observed by W. M. Thornton,† for example, that a spark produced between point electrodes by the discharge of a condenser will ignite a given mixture when another, of the same total heat but produced by separating the electrodes from contact so as to interrupt a current in an inductive circuit, will fail to do so. Experiments of this nature show that the total heat of a spark is not, in general, the property which determines the igniting power, and some observers have,

---

* The greater part of the substance of this chapter was contained in a lecture to Section A of the British Association at Glasgow on 10th September, 1928. See also *Phil. Mag.*, 6, p. 1090 (1928).

† *Phil. Mag.*, (November, 1914).

consequently, been led to assume that the igniting power is determined not by any of the thermal effects but by other physical properties of the spark. One view which has received considerable support is that ignition depends upon the amount of ionization in the gas, and, therefore, upon the current flowing in the spark.

Influence of Duration of Discharge. Another discovery, however, made by J. D. Morgan,* led to a re-examination of the "thermal" theory from a new point of view. In experimenting upon ignition by high-tension sparks (i.e. the sparks from a magneto or an induction coil), Morgan found strong reasons for believing that the ignition in some cases was caused entirely by the initial part of the discharge. We have seen in Chapter V that the discharge produced by such high-tension apparatus consists of two parts, the initial portion representing the discharge of the static electricity accumulated on the secondary coil and conductors connected with it, and the subsequent portion forming an arc discharge in the conducting path prepared by the earlier portion. These two parts of the discharge are of very different duration, the first occupying a very short time, the arc much more prolonged, and the question presented itself whether a source of heat which is active for an extremely short interval of time is likely to be more effective in ignition than one of the same total heat but of greater duration in time.

Minimum Initial Volume of Flame. Some calculations,† based on the assumption that ignition depends upon the *volume* of the gas which the source can by its own heat raise to the ignition temperature,‡ showed that there is good reason for believing that this is the case. The reason for this assumption is as follows: If we suppose that a small spherical volume of the gas is heated by the source to the ignition temperature, the gas within this volume is burnt and its temperature is raised further by the heat resulting from the chemical action. At the surface of the sphere there will, therefore, be a large temperature gradient and rapid loss of heat by conduction.

* *Principles of Electric Spark Ignition*, p. 28. See also Paterson and Campbell, *Proc. Phys. Soc.*, p. 177, 1919.

† *Phil. Mag.*, 43, p. 359 (1922).

‡ See Wheeler, *Trans. Chem. Soc.*, 117, p. 903 (1920); also the "Third Report of the Explosions in Mines Committee of the Home Office" (1913).

The rate of cooling of the sphere due to this cause is proportional to the ratio of its surface to its volume, and is very great if the sphere is very small. Consequently, the small flame started in the sphere will soon become extinguished by the conduction from its surface, and will, therefore, fail to spread throughout the gas, unless the volume of the sphere exceeds a certain minimum value.

Theorem of Point Sources. A condenser spark of very short length between metal points being regarded as an instantaneous point source of heat in a uniform medium, the temperature θ in its neighbourhood is represented by Fourier's expression*—

$$\theta = \frac{Qe^{-r^2/4kt}}{8c(\pi kt)^{3/2}} \qquad . \qquad . \qquad . \qquad . \qquad . \qquad . \qquad . \quad (120)$$

in which Q is the quantity of heat dissipated in the spark, k is the thermometric conductivity, and c the thermal capacity per unit volume of the gas (supposed uniform), r is the distance from the source, and t is the time after the moment at which the heat is communicated. If an inductance spark be regarded as a source in which the heat is supplied to the gas at a uniform rate over an interval of time T, the temperature distribution may be deduced from (120) by integration. In the paper cited the results of numerical calculations based upon (120) were given, which showed that in the case considered the volume of the spherical portion of the gas, the boundary of which was just raised to a certain temperature, was greater in the case of the instantaneous source than in that of a source in which the heat supply was continued at a uniform rate for a finite interval of time, the total heat supplied being the same in both cases.

General Proof of Theorem. The general proof that this result holds also for a point source in which the heat Q is supplied over a finite interval of time, whether uniformly or not, may be arrived at in the following manner :—In Fig. 104 let curve A†️ represent the form of the temperature wave (θ, t) at any distance r from an instantaneous point source. The temperature

---

* *Théorie de la Chaleur*, Sect. 385.

†️ Curve A in Fig. 104 is calculated from the expression (120) with $Q = 0.001$ calorie, $r = 0.0604$ cm., $k = 0.2188$, $c = 0.00032$. The maximum temperature is $1044°$ C. at $t = 0.002779$ second.

at this distance rises rapidly to a maximum and falls more slowly from it. The maximum temperature is attained at the time $t = r^2/6k$, and is higher the shorter the distance r from the source, being, in fact, inversely proportional to the cube of the distance from the source, as may be seen by substituting this value for t in (120).

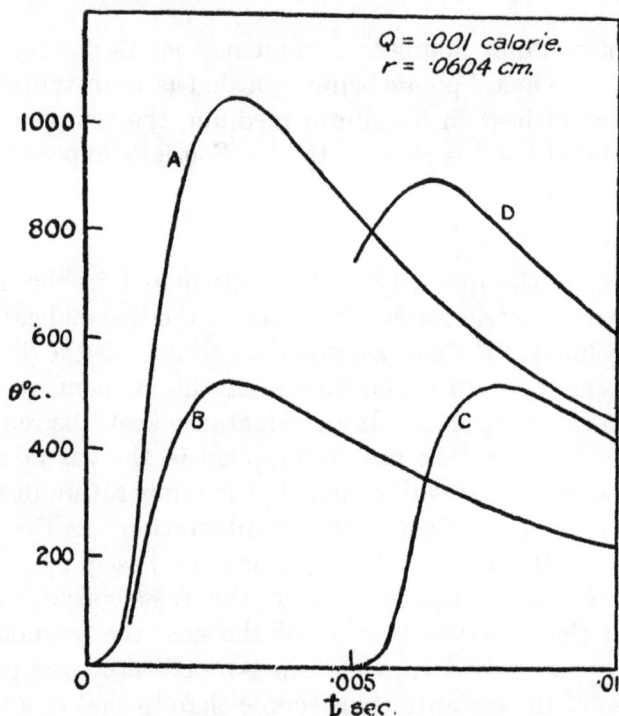

FIG. 104. TEMPERATURE WAVE FORMS DUE TO INSTANTANEOUS AND CONTINUED POINT SOURCES OF HEAT

If we now suppose that the heat Q is communicated in two equal parts at an interval of time T (represented in Fig. 104 by 0·005 sec.), the temperature at the same distance is given by the sum of the ordinates of the two curves B and C, each of which has one-half the amplitude of A. The maximum in the resultant curve occurs at a time shortly before the maximum of the second component, and it is evident that the resultant maximum is smaller than the sum of the maxima of the two components, and therefore than the maximum of the original curve A. The resultant maximum also evidently diminishes as the interval of time between the two components increases. We may conclude that the result of dividing the heat supplied into two equal instalments separated by any finite interval of time is to lower the maximum temperature at any given point in the neighbourhood. Similar considerations show that the same result holds if the two instalments are unequal, also if the heat is divided into three or more instalments, equal or unequal, supplied at equal or unequal intervals of time. The limiting case of a continued source, i.e. a very large number of infinitesimal instalments following one another at infinitely short intervals of time, is also included. Curve D in Fig. 104 shows a portion of the temperature wave at the same distance from a point source of the same total heat, but in which the heat supply is continued uniformly for 0·005 second.

The general result may be stated as follows: If a given quantity of heat is supplied at a point of a uniform conducting medium in any manner during a finite interval of time, the maximum temperature at any neighbouring point is lower than it would have been if the heat had been supplied all at the same instant.

By considering the distance from the source at which the temperature just rises to a given value, instead of the maximum temperature at a given distance, we arrive at the following corollary to the above theorem: If a given quantity of heat is supplied at a point of a uniform conducting medium in any manner during a finite interval of time, the volume of the medium, the boundary of which is just raised to any given temperature, is smaller than it would have been if the heat had been supplied all at the same instant. This follows from the theorem and the result, previously stated, that for instantaneous sources the maximum temperature diminishes with increasing distance from the source.

The introduction of "volume" instead of "distance" in the corollary follows from the assumed uniformity of flow of heat in all directions. Since, however, the proof of the theorem does not depend upon the precise form of curve A in Fig. 104, the theorem and its corollary are applicable to the case of a point source in a conducting medium between two plane parallel

non-conducting walls at a short distance apart, or to that of
a point source in a thin column of conducting material bounded
laterally by non-conductors. If in these cases the bounding
walls were made conducting some of the heat would enter the
walls and would thus be lost to the medium between them,
but since there seems to be no reason for supposing that the
medium would lose more heat in this way with the instantaneous source than with the divided or continuous source, we

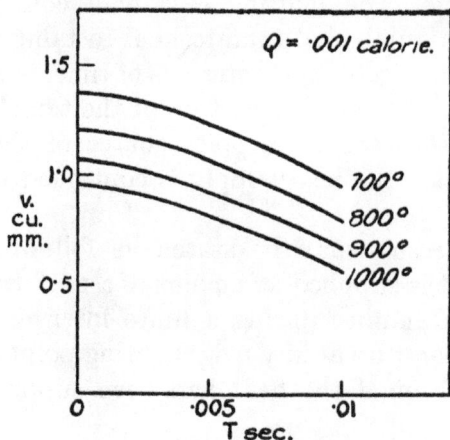

Fig. 105. Effect of Duration of Point Source on Volume
Raised to Ignition Temperature

may assume the theorem and its corollary to hold also in this
case.

Numerical Values. The magnitude of the effect of the duration of the heat supply is illustrated by the curves in Fig. 105,
in which the ordinate represents the volume of gas, the boundary of which is just raised to a definite temperature by a point
source of heat continued at a uniform rate for a time T represented by the abscissa. The volumes are calculated, from the
integral of the expression (120),* for four temperatures (within
the range of the ignition temperatures of methane-air mixtures),
the values assumed for the constants being those given in the
footnote on page 197. It will be seen that the volume is greatest
for an instantaneous source ($T = 0$), and that it diminishes
steadily as the duration of the heat supply is increased. The

* See *Phil. Mag.*, p. 364 (February, 1922).

same holds when the heat supply, instead of being continued uniformly for time T, is divided into a given number of equal instalments supplied at equal intervals over this time. It is the increase in the total duration of the heat supply, rather than an increase in the number of instalments, which causes the reduction in the volume raised to the given temperature. An increase in the number of instalments (supposed equal and equally spaced in time), without increase in the total duration, has the opposite effect.

According to the hypothesis that ignition depends upon the volume of the gas which is raised to the ignition temperature, it follows from the corollary stated above that an instantaneous point source of heat is more effective in ignition than a point source in which the heat is supplied in any manner (continuous or discontinuous) over a finite interval of time.

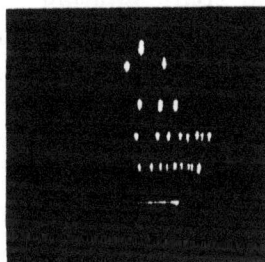

Fig. 106. Sources of Different Time Distribution

Sources of Different Time Distribution. Experimentally it is easy to produce sources of approximately the same total heat, but having different time distributions, by connecting a condenser of variable capacity to the electrodes of a spark-gap connected with the secondary of an induction coil fed with a primary current which has a constant value at the moment of break. With a large value of the secondary capacity a single spark is obtained, and as the capacity is diminished the discharge produced by each interruption of the primary current divides into two, three, or more sparks. When the secondary condenser is disconnected we have the ordinary induction coil discharge consisting of a preliminary spark followed by a continuous but pulsating and decaying arc.

In Fig. 106 are shown six induction coil spark discharges produced in this way, between the ends of two wires at a very short distant apart, and photographed with the aid of a rotating mirror. That the energy dissipated in the six discharges was approximately the same is a consequence of the fact that the primary current at break, and therefore the energy supplied to the system, was the same in all. Calorimeter measurements of the heat of the sparks in such experiments also show that it is practically independent of the secondary capacity.*

An experiment described by Morgan† shows that the effectiveness of a magneto spark in ignition increases with the capacity of a condenser connected with the spark-gap electrodes. This result is in agreement with the theoretical views described above, since the total duration of the discharge in general diminishes as the secondary capacity is increased (see Fig. 106).

Explosion Vessel. A simple form of explosion vessel which the present writer has found suitable for illustrative experiments on spark ignition consists of a strong glass tube of uniform bore, about 16 in. long and $1\frac{1}{4}$ in. internal diameter, slightly constricted at the upper end in order to hold firmly a plug of insulating material which closes the tube at this end and carries a gas inlet-tube and the electrodes of an adjustable spark-gap. A piston of felt, movable easily in the glass cylinder, is fitted to the end of a long brass tube of $\frac{1}{4}$ in. diameter which acts as air inlet. The glass cylinder is fixed firmly in a vertical position, the brass tube passing through a guide about 2 in. below the lower end of the cylinder. The upper surface of the guide also acts as a buffer from which the piston rebounds after the explosion. A scale of inches runs along the length of the cylinder, to measure the air introduced, the uppermost inch being divided into tenths for the measurement of the volume of gas admitted.‡ A magneto (or an induction coil) and a condenser connected with the spark electrodes complete the apparatus. The piston is first set at a suitable mark near the top of the cylinder, gas is admitted for a short time and is then cut off by a stopcock in the gas inlet-tube. The piston is drawn down to a suitable distance (depending upon the strength of mixture required) and a stopcock near the lower end of the air inlet-tube is then closed. A half-turn of the magneto armature (giving one break of the primary circuit)

---

* A form of calorimeter suitable for comparative measurements of the heat of sparks consists of a gas thermometer the bulb of which contains the spark-gap. The deflexion of the liquid column produced by a series of sparks, and measured from the zero observed after sparking has ceased, gives a measure of the heat produced in the sparks.

† *Electric Spark Ignition*, pp. 31, 32 (1920); *Engineering* (3rd Nov., 1916).

‡ Explosion tubes of this form were used for illustrating the lecture in Glasgow.

produces the spark, and if the mixture is such that an explosion of sufficient violence results, the piston is blown out of the cylinder and after the rebound re-enters the cylinder to a certain distance. A scale of tenths of an inch may be marked on the cylinder at the lower end to indicate this distance, which gives a measure of the impulse of the explosion.*

By means of this apparatus the igniting action of sparks of different kinds between electrodes of different forms can be conveniently studied, and the superiority of a condenser spark over the ordinary magneto spark (e.g. with pointed electrodes of steel or tungsten), indicated by the above theory, can easily be demonstrated.

Ignition by Sparks between Spherical Electrodes. When ignition experiments are tried with short sparks between spherical electrodes of metal † a number of results are found which appear, at first sight, to be contrary to the thermal theory. In the first place ignition is difficult and very erratic unless the metal surfaces are clean. It is scarcely to be expected, however, that the loss of heat which undoubtedly takes place by conduction from the gas to the electrodes would be less when these are clean than when they are covered with a layer of oxide or other substance of smaller conductivity than the metal. The result, therefore, seems to point to some action other than thermal as the cause of ignition. Further, when different metals are used as electrodes, the igniting effect does not seem to depend appreciably upon their thermal conductivity. In making this comparison care should be taken to use electrodes of the same curvature (since the igniting power of the spark increases with their curvature), and they should be well cleaned before the experiment. In this way curved surfaces of copper, steel, and zinc were found to be equally effective in ignition, i.e. to be just capable of igniting a given mixture when the sparks were of the same kind and length, and were produced by the interruption of the same primary

---

* The pressure developed in an explosion is usually independent of the nature and intensity of the spark, so long as the spark is sufficient to produce an explosion at all (see Morgan, *Electric Spark Ignition*, p. 15).

† Short cylinders of about 8 mm. diameter, having spherical ends of about 1 cm. radius of curvature, and placed so as to form a gap 0·15 mm. wide at its narrowest part, were used in most of the experiments described in this section.

current. Of the metals examined (copper, steel, zinc, platinum, lead) lead was found to be the most suitable for ignition experiments; the surface of this metal is less easily spoilt by the tarnishing due to the sparks or to the flame. Carbon electrodes, though not so effective in ignition as those of clean lead, are also suitable because they do not require to be cleaned as do metallic surfaces.

Another result which appears to be contrary to the thermal theory is found when the effect of connecting a condenser to the spherical electrodes is examined. With electrodes and gap of the dimensions stated above, the effect of the condenser is exactly opposite to that observed when pointed electrodes are used. The condenser produces a decided diminution in the igniting power of the spark, and the inferiority of the condenser spark with the spherical electrodes is quite as marked as its superiority when the electrodes are metal points. In one experiment, with spherical carbon electrodes, ignition without the condenser occurred at a primary current of 0·7 amp.; with the condenser ignition failed at 10 amp., i.e. with a spark of nearly 200 times as much energy.

This result is directly contrary to that derived from the theory of thermal conduction from point sources, and we must conclude either that the thermal theory is wrong or that some other action takes place, when spherical electrodes are employed, which is of such greater influence in ignition than thermal conduction as to mask its effect.

Spreading of Discharge over Surfaces of Electrodes. For the further investigation of this matter some photographs of the sparks between spherical electrodes were taken, and six reproductions are shown in Fig. 107. These sparks were produced, by a magneto, between lead cylinders 8 mm. in diameter, the spherical ends of which were set at 0·15 mm. apart. The camera used in photographing them had a quartz lens of 15·6 cm. focal length, the linear magnification being 1·5. The sparks shown in Fig. 107 are "ordinary" magneto sparks, no condenser being connected with the electrodes.

An examination of Fig. 107 shows that the ordinary spark between spherical electrodes differs in one important particular from the usual short spark which we have regarded as a point source. The discharge begins at the centre (i.e. the narrowest part) of the gap, but some portion of it spreads towards the sides, and in spreading it lengthens so that it can no longer be regarded as even approximately a point source. That the

Fig. 107. Spreading Discharges Between Spherical Electrodes

spreading is not an effect caused by gaseous combustion is shown by the fact that the discharges in Fig. 107 took place in ordinary air free from inflammable gas. The spreading of the discharge over the electrode surfaces does not occur when a condenser of considerable capacity is connected directly with

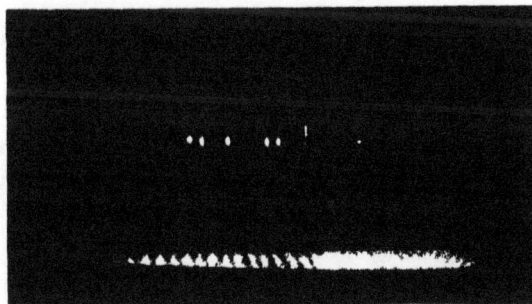

Fig. 108. Ignition Produced only by the Spreading Discharge

them. In Fig. 108 are shown seven induction coil sparks between the lead cylinders, the gap in each case being placed centrally just above a Meker burner.* The first five passed

* Ignition experiments with a burner may be made either while the gas is flowing or in the still gas which remains above the burner for a short time after the gas is turned off. A large Meker burner is the most suitable for the purpose.

while a condenser was connected with the terminals; they show no spreading and they failed to ignite the gas.* The sixth and seventh sparks were produced after the condenser had been disconnected; of these the sixth shows spreading and produced ignition, the seventh shows no spreading and failed to ignite. The primary current interrupted was the same in all seven.

In Fig. 109 are shown seven magneto sparks between spherical electrodes of platinum. Of the seven only the second and

FIG. 109. IGNITION ONLY BY THE TWO SPREADING DISCHARGES

the fifth show evidence of spreading, and only these two produced ignition.

Conclusions from Examination of Photographs. The conclusions to be drawn from an examination of these photographs, and a number of others of a similar kind which were taken, are that, in the case of short sparks between spherical electrodes of metal or of carbon—

1. The ordinary high-tension discharge (without secondary condenser) tends to spread from the centre towards the sides of the gap when the electrodes are of carbon or of clean metal.

2. The tendency to spread increases with the primary current.

3. The spreading does not occur, or occurs much less readily, over metal surfaces which are not clean.

4. Ignition does not occur, or occurs only with great difficulty, unless the discharge spreads.

* Owing to the much greater brightness of the condenser sparks the aperture of the lens was reduced to its minimum for the first five sparks. Their images are, however, still rather enlarged by halation.

5. Ignition does not always occur if there is spreading.

6. Spreading does not occur if a condenser of considerable capacity* is connected directly with the electrodes.

Results with Spherical Electrodes. With the foregoing in mind it is easy to understand why the ordinary spark is a better igniter than the condenser spark when spherical electrodes are used. The ordinary discharge, in spreading to the outer and wider parts of the gap, is able to warm the requisite volume of gas to the ignition temperature, not by thermal conduction but by its own expansion. The condenser spark, on the other hand, being confined to the narrowest part of the gap, can warm the surrounding gas only by conduction.† It is true that in the case of the ordinary spark between spherical electrodes, only a fraction of the heat is actually utilized in producing ignition, viz. the heat of that portion of the discharge which is near the edge of the gap. It is quite in accordance with the thermal theory, however, that an enlarged source may be a better igniter than a point source of greater heat. For the distribution of temperature round an instantaneous point source at any time after the heat is communicated is such that the temperature is highest in the position of the source, and falls off in all directions from this point. If the boundary of a certain volume of the gas is at the ignition temperature, the inner portions of this volume must be at a temperature above this, and therefore at an unnecessarily high temperature for the production of ignition. It is clear that, in regard to the volume raised to the ignition temperature, a better distribution would be one in which the heat is more evenly distributed, so that the temperature throughout this volume is uniform. An instantaneous point source, though superior to a continued point source, is inferior to an enlarged source of the same or even less total heat.

It appears, therefore, that the results, both with pointed

---

* The capacity must be sufficiently large to prevent the formation of an arc instead of a single or multiple arc (see p. 130).

† It might be expected that the condenser spark, being in the position in which it can give heat most readily to the electrodes, would communicate less heat to the gas than would the ordinary spark. Calorimeter experiments, however, with small spherical electrodes of carbon indicated that the gas receives slightly more heat from condenser sparks than from ordinary sparks produced with the same primary current.

and with spherical electrodes, are consistent with the view that the most effective spark in ignition is that which heats the greatest volume of the gas to the ignition temperature. With pointed electrodes the heating is effected by thermal conduction from the source, with spherical electrodes by expansion of the source itself.

Wandering of the Arc. Some further points now remain to be considered in regard to the discharge between spherical

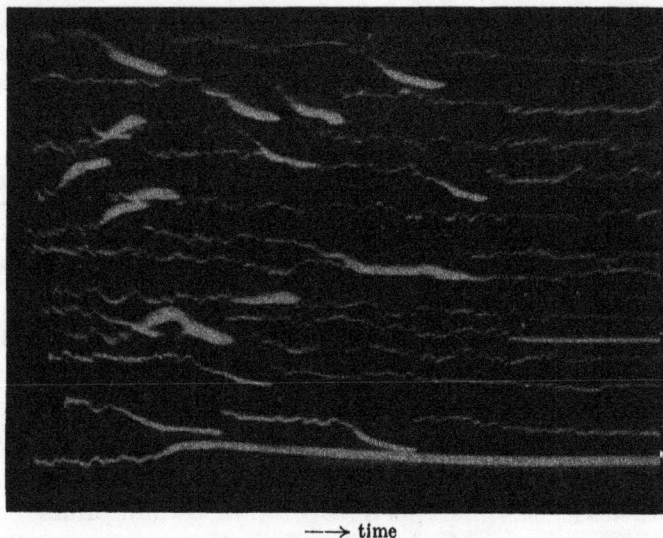

———> time

FIG. 110. WANDERING ARCS BETWEEN SPHERICAL ELECTRODES
(Beginning of discharges at left-hand side.)

electrodes. The horizontal striations which appear in the photograph of the spreading discharge in Fig. 108, and less clearly in Figs. 107 and 109, suggest that the apparent spreading is a radial movement, or wandering, of the arc portion of the discharge, the striations corresponding to the oscillations of the induction coil or magneto system.* That this is the case is confirmed by photographs of the discharge taken with the help of a rotating concave mirror, some of which are reproduced in Fig. 110. They show a number of spark discharges, produced by an induction coil without secondary condenser, between spherical electrodes of carbon. Nearly all the images

* See Fig. 61, p. 122, and Fig. 103, p. 193.

show the wandering of the arc, the movement being upwards, or downwards, or along other radii. Frequently the arc wanders to the edge of the gap and breaks off there, the discharge then beginning again at the centre, sometimes afterwards wandering along a different radius, as in the fourth, fifth, and sixth images. This is the explanation of the fact that in several of the camera photographs (e.g. the first in Fig. 107) the "spreading" appears to take place both upwards and downwards in the same discharge. None of the lines in Fig. 110 show any bifurcation, the discharge passing at only one part of the gap at a time. The curvature of the lines during wandering shows that the speed of the lateral movement of the discharge is greatest at the centre of the gap, which is to be expected, since the radial variation of the width of the gap is here smallest.*

With regard to the influence of the wandering on ignition, it is probable that the most effective discharges in this respect are those in which the arc wanders to the edge of the gap and remains there for an appreciable time, as in the sixth and thirteenth images in Fig. 110. On several occasions it was observed that a discharge which broke off just after reaching the edge, to recommence at the centre, was incapable of producing ignition. In such cases it must be concluded that though the wandering is accompanied by a sufficient enlargement of the source, the time for which the enlargement endures is too short to result in ignition.†

The wandering of the arc takes place less readily when the width of the gap is increased. Consequently, it might be expected that within certain limits a narrow gap between spherical electrodes would be more effective in ignition than a wider one. This was found to be the case with the carbon spherical electrodes used in the present experiments. The igniting effect of the spark was distinctly better when they were 0·15 mm. than when they were 0·3 apart, the heat of the spark being practically the same on both occasions. The

---

* The photographs reproduced in Fig. 110 suggest that the wandering is accompanied by an increase of width, as well as an increase of length, of the arc.

† The whole time of duration of each of the discharges shown in Fig. 110 was about 1/100 second.

greater ease of wandering in the narrower gap was more effective than the greater length of the initial spark in the wider one.

The wandering also depends upon the curvature of the electrodes, and in the case of sharply-pointed electrodes it must be greatly restricted by the fact that here any lateral movement of the arc would be accompanied by excessive increase of length. The possibility suggested itself, however, that some slight effect of wandering might be observable with pointed electrodes if these were of the most suitable material. When ignition was tried with a very short spark between carbon points it was found that the addition of a secondary condenser now produced no improvement. Ignition was effected with equal success whether the condenser was connected or not. The same was found with pointed electrodes of clean lead. In these circumstances the slight wandering over the sides of the electrodes in the case of the ordinary spark apparently produces as much effect in ignition as the superior temperature distribution due to thermal conduction in that of the condenser spark.

Cause of Wandering of the Arc. As to the cause of the wandering of the arc, this cannot be traced to the action of thermal convection arising from the heating of the gas by the initial spark. The photographs reproduced in Figs. 107 and 110 show that the movement of the arc is as often downwards as upwards. Nor can the wandering be attributed to thermionic action, since the movement is from the centre towards the outer portions of the gap where the surfaces of the electrodes are cooler. The wandering must be attributed to some property of the surfaces which is independent of thermal action. Now it is an observed fact that the wandering takes place much less readily if the electrode surfaces are not clean or are tarnished, and this fact suggests that photoelectric action plays a part in determining the position of the arc. We may suppose that the electrode surfaces at the centre are to some extent "spoilt" by the initial spark, so that the arc which follows it passes more readily across a neighbouring part of the gap where the surfaces are cleaner. The wandering of the arc thus represents the tendency of the arc to pass across parts of the gap where the surfaces have not been spoilt by the previous portions of the discharge. According to this view of the matter the direction of the wandering, which is apparently quite capricious, is that along which the surfaces at the time are cleanest, and where the easiest path is prepared for the discharge by photoelectric action.

In Fig. 111 are reproduced rotating mirror photographs of four multiple spark discharges produced between the carbon

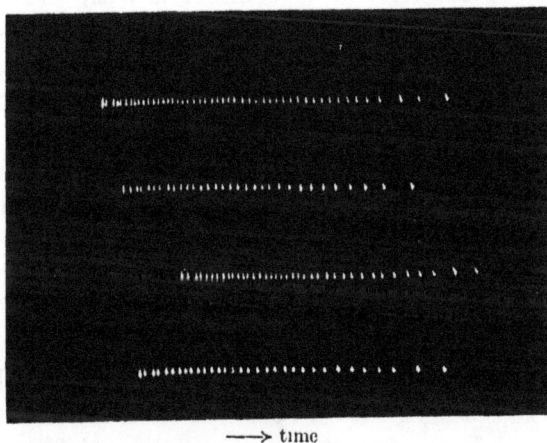

$\longrightarrow$ time

Fig. 111. Multiple Sparks Between Spherical Electrodes

spherical electrodes by an induction coil with secondary condenser. The capacity of the condenser was considerably less than the maximum which allowed sparks to pass, so that each discharge consisted of a large number of separate sparks. Each spark appears at the centre of the gap, and no part of the discharge shows any tendency to wander towards the side. This is probably due to the short duration of the high-frequency oscillations in each of these sparks. If the frequency is lowered by the inclusion of an inductance coil in the condenser-gap circuit, the sparks show some evidence of a tendency to wander.*

A Test of the Electrical Theory of Ignition. It has already been mentioned that, according to the electrical theory, it is

---

* It was found by G. I. Finch and H. H. Thompson (*Proc. Roy. Soc.*, A. 134, p. 343, 1931) that the igniting power of a high frequency spark increased when the frequency was lowered in this way.

not any thermal action of the current but the value of the current itself which is the determining factor in ignition by electric sparks or other forms of electric discharge. It is not difficult to show, however, that in the present experiments with spherical electrodes the maximum current crossing the spark gap is much greater in the case of the condenser spark than in that of the ordinary spark, though the latter is, as we have seen, much the better igniter. Let us suppose that a condenser of capacity C is connected with the electrodes, so that the discharge takes the form of a single spark in which the current oscillates with high frequency n, determined by the capacity C, and the self-inductance of the short wires by which it is connected to the electrodes. Then if the sparking potential is V_0, the maximum value of the current in these oscillations is (if we neglect damping) $2\pi n C V_0$. In the present experiments V_0 was about 1,000 volts, C about 0·004 mfd., and n was not less than 10^6 per sec. Consequently, the maximum current in the condenser spark was, at a low estimate, 25 amp.

When the condenser is replaced by one of very small capacity, the discharge changes into a spark of the ordinary kind, consisting of an initial capacity portion followed by a pulsating and decaying arc. The maximum current in the capacity portion is given by the same expression with the appropriate values of n and C, and since n is inversely proportioned to $\sqrt{C}$, the maximum current is now considerably less than before, being directly proportional to the square root of the capacity.

As to the maximum current in the arc portion of the ordinary discharge (in which the current wave consists of one oscillatory and one aperiodic component), an upper limit to its value may be obtained by calculation from the constants of the magneto circuits and the primary current at the moment of "break" (see Chapter V). By such calculations it can be verified that in no case does the maximum current in the inductance portion of the spark given by a high-tension magneto of the usual type exceed a few hundred milliamperes.* It is, therefore, quite

---

* See also Morgan, *Electric Spark Ignition*, p. 21, where it is shown that the current in the arc portion is smaller than that in the capacity portion of a magneto spark.

certain that when a condenser of considerable capacity is connected with the spark electrodes the maximum current in the discharge is much greater than that in the discharge which occurs when the condenser is replaced by one of very small capacity, or when the condenser is absent. The great superiority of the igniting action of the ordinary discharge over that of the condenser spark between spherical electrodes cannot therefore be traced to any direct electrical action determined by the value of the current.

Interpretation of Thermal Theory of Ignition. Although the thermal theory has proved to be capable of reconciling all the facts known about ignition by electric sparks, so far as they have been compared with it, it should be borne in mind that the theory does not offer any explanation of the mechanism of ignition. There is nothing in the thermal theory, for example, to suggest that the energy of the translational motion of the molecules, or that of their rotational or vibratory motion, plays a preponderating part in the process of ignition. Each of these portions of the molecular energy increases with rise of temperature, and all that the thermal theory states is that if a certain minimum volume characteristic of each explosive gaseous mixture is raised to the ignition temperature, the flame so formed will spread throughout the gas. The importance of thermal conduction in the theory is that it enables the "minimum volume" to be raised to the required temperature more easily with some sources of heat than with others, but it is not suggested that thermal conduction, or the translational energy, plays any particular part in the chemical reactions which occur when that temperature is reached.

The thermal theory is primarily a theory of the initiation of gaseous explosions by small sources of heat, but it does not suggest that a point source of very high temperature is a better igniter than a source of greater volume, but of the same total heat and, therefore, of lower temperature. The most effective source, in fact, for a given heat supply, is that in which the heat is communicated instantaneously to a volume of the gas which it can just raise to the ignition temperature.

As to the existence of a "minimum volume" in ignition by heat sources there is direct experimental evidence. Messrs.

Coward and Meiter* have measured the least volume which must be heated, by a condenser spark in a methane-air mixture, to the ignition temperature in order to ensure general inflammation, and have found that the volume is of the order of magnitude indicated by calculation from the expression (120). Also, in some recent experiments on ignition by small flames,† Mr. J. M. Holm has been able to produce within an explosive gas mixture a spherical flame so small that it remains quiescent for a considerable time, whereas a slightly larger flame immediately produces an explosion. The flame of minimum volume, or rather the maximum that can exist without expanding, can, therefore, be seen and photographed.

As already indicated, the above remarks apply more especially to initial flames which are spherical, or nearly spherical, in shape, and if sources of other forms are considered the question arises whether it is always the volume that determines whether the flame will spread. It is possible that in the case of linear or laminar sources, the initiation of an explosion depends upon one of the linear dimensions of the source rather than upon its volume.

* *Journ. Amer. Chem. Soc.*, 49, p. 396 (1927).
† *Phil. Mag.* 1932.

CHAPTER IX

TRIODE OSCILLATIONS IN COUPLED CIRCUITS*

In the previous chapters we have confined our attention mainly to problems in which transient electrical oscillations are set going in coupled systems by causing some sudden change (such as interrupting a battery current or discharging a condenser) in the primary circuit, but at the present time induction coils (the term being used in its wider sense) are very frequently used in circuits in which oscillations are maintained, the most familiar method of maintenance being that depending upon the use of a triode valve. Accordingly, we will in this chapter consider certain matters connected with the maintenance of triode oscillations in coupled circuits and certain transient phenomena which occur when a coupled system is switched into the circuit of a valve oscillator.

The diagram in Fig. 112 shows one of the arrangements of apparatus suitable for experiments on this subject. A and B are the generating coils, connected in the anode and grid circuits respectively, C_1 is the primary condenser, O_1 a current oscillograph, H a hot-wire ammeter, T the transformer or induction coil which may be short-circuited by the key K. O_2 is an electrostatic oscillograph connected with the secondary terminals of the transformer. The deflexion shown by O_1 is proportional to the current flowing through the instrument, that of O_2 to the square of the difference of potential at its terminals. A 60-watt transmitting valve, giving anode currents up to 100 milliamperes, is a suitable generator for the experiments. The short-circuiting key K may be a mercury dipper interrupter actuated by the rotating mirror used with the oscillographs.

The oscillatory system thus consists of two coupled circuits of which the primary includes the coil A, the condenser C_1, the ammeter and oscillograph O_1, and the primary coil of the transformer. The capacity in the secondary circuit includes that of the oscillograph O_2 and spark-gap electrodes, and the

* *Phil. Mag.*, 47, p. 625 (1924).

distributed capacity of the coil, and may be increased by the
addition of a variable condenser connected to the secondary
terminals. This system possesses two frequencies of oscillation
which depend upon the inductances and capacities and upon
the coupling of the two circuits. The coupling is less than

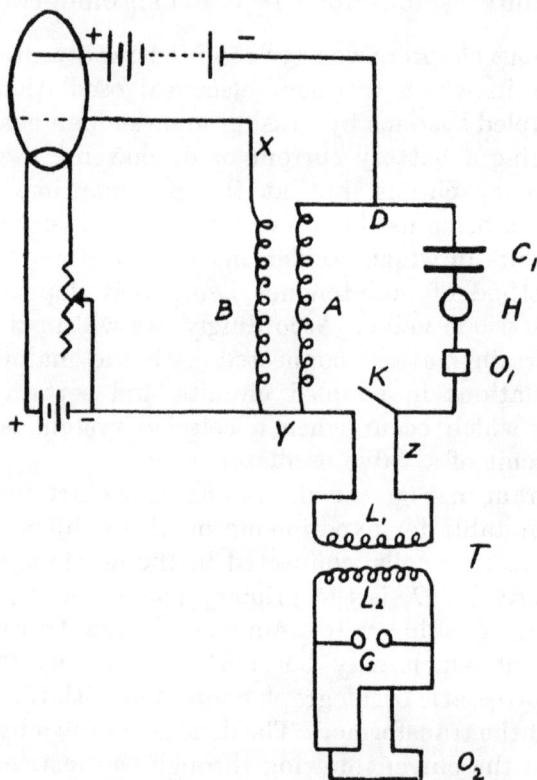

FIG. 112. CIRCUIT OF TRIODE OSCILLATOR WITH TRANSFORMER

$A B$ = Generating coils.	T = Transformer.
C_1 = Condenser.	K = Key.
O_1 = Current oscillograph.	G = Spark gap.
H = Hot-wire ammeter.	O_2 = Electrostatic oscillograph

that of the transformer alone in the ratio of L_1, the self-inductance of the primary coil, to $L_1 + L_A$, the total self-inductance
in the primary circuit. Instead of the connections shown in
Fig. 112, the condenser and transformer, with the ammeter
and oscillograph, may be connected between the points D and
X, in which case the coupling is further reduced owing to the
addition of the coil B to the primary circuit.

The oscillation may be rendered audible by means of a telephone connected with a loop of wire loosely coupled with coil A.

The Phenomena Observed. The following are the effects which we shall consider—

1. When the system is oscillating with the transformer in circuit, one note only is usually heard in the telephone, not two notes simultaneously as might be expected. A similar preference for one or other of the two possible vibrations is also observed in other coupled systems, electrical and mechanical (e.g. the singing arc with coupled circuits, and reed pipes), which give sustained oscillations.

2. If the system is oscillating with the key K closed and the transformer is then thrown into the circuit, the pitch of the telephone note may rise, or fall, or it may undergo no change. These various results may be produced by suitably adjusting the capacity of the condenser C_1.

3. If the capacity C_1 is gradually varied with the induction coil always in circuit, the pitch of the note changes suddenly at a certain stage. This abrupt change of pitch may take place directly or with an interval in the sequence of values of C_1 over which no oscillation is produced.

4. In the secondary circuit the potential produced at unshortcircuiting is generally much higher than that generated afterwards by the alternating current flowing in the transformer. In fact, the valve may be unable to maintain any oscillation with the transformer in circuit, and there may still be a high potential generated at the secondary terminals when the transformer is switched in.

Some of these effects are illustrated by the following current oscillograms. Fig. 113 is a typical curve showing the effect of throwing the transformer into circuit. In this experiment the pitch of the note rose by about a major third at unshortcircuiting, and the sound was, at the same time, reduced in intensity, as indicated by the small amplitude of the maintained oscillations in the latter part of the curve. Fig. 114 shows a similar effect but with a greater rise of pitch (in this case a minor sixth) at the introduction of the transformer.

In the experiments of Figs. 113 and 114, no spark appeared at the secondary terminals either at unshortcircuiting or afterwards. If the gap is reduced so as to allow a spark to appear at unshortcircuiting, the amplitude of the curve falls much more

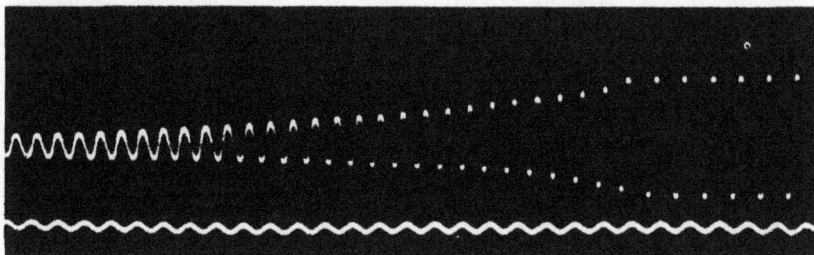

FIG. 113. CURRENT OSCILLOGRAM SHOWING INCREASE OF FREQUENCY AND DIMINUTION OF AMPLITUDE ON SWITCHING IN TRANSFORMER

rapidly to a smaller value, and afterwards declines slowly to its final value, as shown in Fig. 115 (8·3 mm. spark). With a shorter gap (3·4 mm.) the final steady state is reached sooner, the shorter spark evidently taking more energy from the system

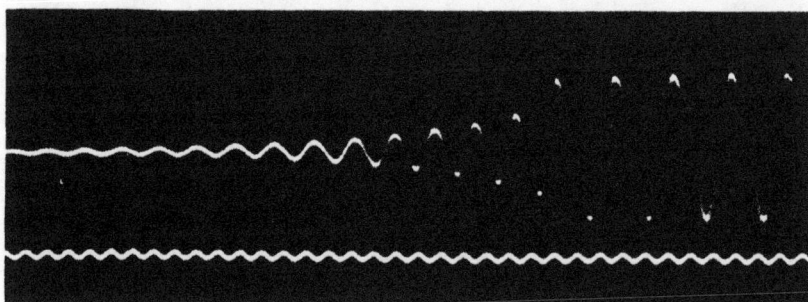

FIG. 114. SHOWING LARGE INCREASE OF FREQUENCY ON SWITCHING IN TRANSFORMER

(Fig. 116). If the gap is still further reduced (or the current increased) so that sparking takes place freely under the maintained alternating current flowing in the transformer, the curve is as shown in Fig. 117. It appears that each spark has the effect of reducing the amplitude of the oscillation, and that some time is required for it to recover and grow again to the value needed for a repetition of the discharge.

If the oscillograph, instead of being in the condenser circuit,

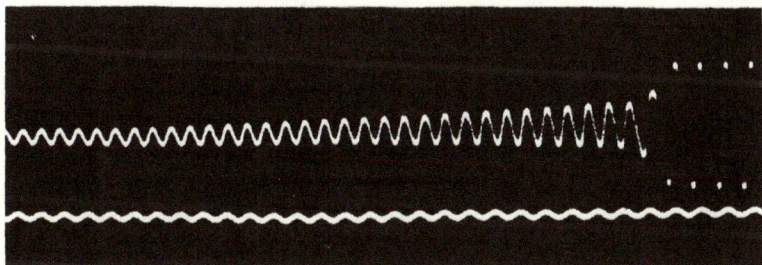

FIG. 115. CURRENT VARIATION ON SWITCHING IN TRANSFORMER
WITH SPARK

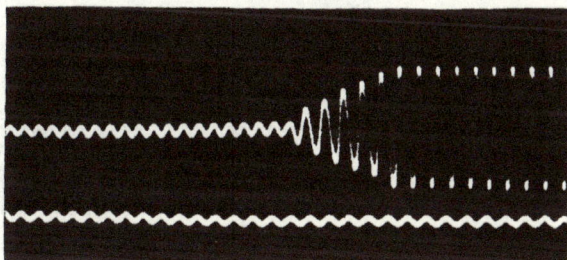

FIG. 116. TRANSFORMER SWITCHED IN WITH SHORT SPARK

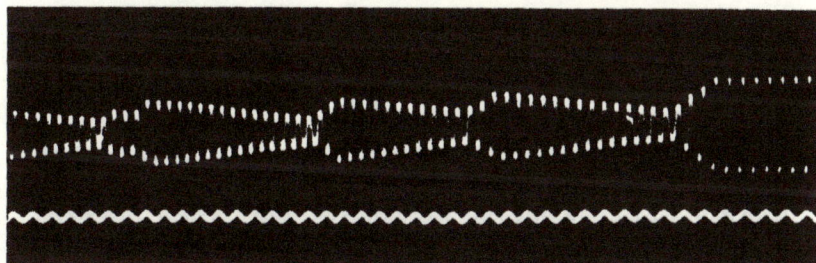

FIG. 117. SERIES OF SPARKS AFTER TRANSFORMER IS ADMITTED
TO CIRCUIT

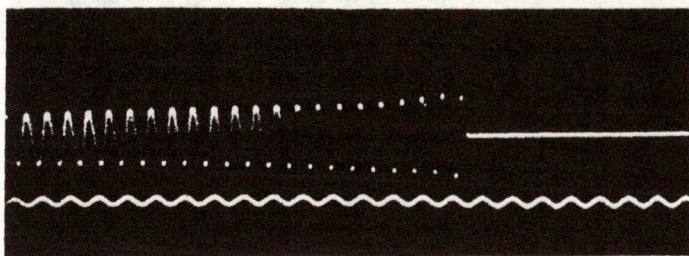

FIG. 118. OSCILLOGRAM OF CURRENT IN TRANSFORMER CIRCUIT

is connected in the transformer primary circuit, as at Z (Fig. 112), the instrument will, of course, show only the current flowing after the key K is opened (Fig. 118).

The curve of Fig. 119 was taken with an induction coil

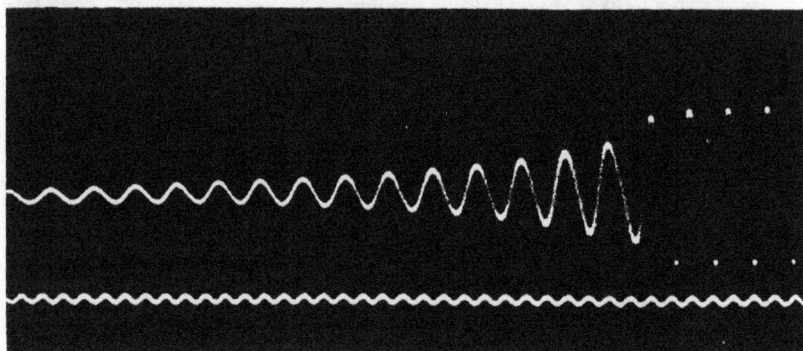

FIG. 119. SHOWING FALL OF FREQUENCY ON SWITCHING IN INDUCTION COIL

substituted for the transformer. On this occasion the condenser C_1 (2·63 mfd.) with the induction coil and oscillograph were connected between D and X (Fig. 112). The curve indicates a fall of pitch of about a semitone at "unshortcircuiting."

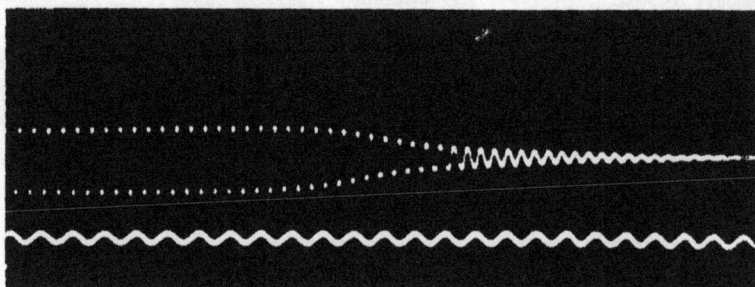

FIG. 120. SHOWING BEGINNING OF OSCILLATIONS

Fig. 120 shows the beginning of a train of oscillations (condenser, coil, and oscillograph between D and Y as in Fig. 112) produced by closing a contact maker in the grid circuit.

It will be seen that all these curves show a single oscillation of increasing, diminishing, or constant amplitude when the transformer or induction coil is in circuit and not sparking, and that there is no evidence of a periodic fluctuation of amplitude such as might be expected in a system possessing two frequencies of oscillation.

We will now proceed to consider the reasons for the peculiar changes of pitch at unshortcircuiting, and for the almost complete absence of any trace of one or other of the two component oscillations of the system, both in the telephone sound and in the current oscillograph curves.

The Apparent Self-inductance and Resistance of the Primary Coil of a Transformer. In the first place it is necessary to consider certain points in the theory of the transformer regarded as a coupled system having electrostatic capacity in the secondary circuit, to the primary of which an alternating potential of given frequency and constant amplitude is applied. Since the secondary capacity is in part distributed over the coil, and, in consequence, the current in the secondary wire varies along its length, we shall, as before, denote by i_2 the current in the central winding of the secondary coil, and by L_{21}, L_{12}, the coefficients of induction between the primary and secondary. If the E.M.F. applied to the primary terminals has amplitude E and frequency $p/2\pi$, the equations for the primary and secondary currents are—

$$L_1\frac{di_1}{dt} + L_{12}\frac{di_2}{dt} + R_1 i_1 = Ee^{ipt} \qquad . \qquad . \qquad . \qquad . \quad (121)$$

$$L_2\frac{di_2}{dt} + L_{21}\frac{di_1}{dt} + R_2 i_2 + V_2 = 0 \qquad . \qquad . \qquad . \quad (122)$$

Where V_2 is the potential at the secondary terminals. Also, if C_2 is the secondary capacity,

$$i_2 = C_2\frac{dV_2}{dt} \qquad . \qquad . \qquad . \qquad . \qquad . \quad (123)$$

Assuming that in the forced oscillation i_1 and i_2 vary as e^{ipt}, and eliminating i_2 and V_2, we arrive after reduction at the equation for i_1,

$$L\frac{di_1}{dt} + R i_1 = Ee^{ipt} \qquad . \qquad . \qquad . \quad (124)$$

where
$$L = L_1 + \frac{L_{21}L_{12}p^2C_2(1 - p^2L_2C_2)}{(1 - p^2L_2C_2)^2 + R_2^2C_2^2p^2}* \qquad . \qquad . \quad (125)$$

$$R = R_1 + \frac{L_{21}L_{12}p^4C_2^2R_2}{(1 - p^2L_2C_2)^2 + R_2^2C_2^2p^2}* \qquad . \qquad . \quad (126)$$

Introducing the coupling of the transformer k^2 ($= L_{21}L_{12}/L_1L_2$), and denoting $p^2L_2C_2$, the ratio of the squares of the applied frequency and the natural frequency of the secondary circuit when oscillating alone, by f, and the quantity $\left(\frac{R_2}{L_2}\right)^2 L_2C_2$ by λ, we find for L and R the expressions—

$$L = L_1\left[1 + \frac{k^2f(1 - f)}{(1 - f)^2 + \lambda f}\right] \qquad . \qquad . \qquad . \quad (127)$$

$$R = R_1\left[1 + \frac{k^2\frac{L_1}{R_1}\frac{R_2}{L_2}f^2}{(1 - f)^2 + \lambda f}\right] \qquad . \qquad . \quad (128)$$

L and R may be called respectively the "apparent self-inductance" and the "apparent resistance" of the primary coil of the transformer. They are the self-inductance and resistance which a single coil would require to have in order to allow, under the action of the applied E.M.F. $E\cos pt$, a current i_1 to pass identical in amplitude and phase with the primary current of the transformer. The apparent impedance of the primary coil is $\sqrt{R^2 + L^2p^2}$, and the angle by which the primary current is behind the applied E.M.F. in phase is $\tan^{-1} Lp/R$.

It appears from the expression (127) that L is equal to L_1 when the applied E.M.F. is varying infinitely slowly ($f = 0$), and again in the case of resonance ($f = 1$). For infinitely rapid variations ($f = \infty$) L is equal to $L_1(1 - k^2)$.

Variation of Apparent Self-Inductance with Frequency. The manner of variation of L with the applied frequency may be illustrated by a numerical example. In one of the experiments the coupling k^2 of the induction coil was 0·768, and the value of λ was 0·005147. With these values the variation of the ratio L/L_1 with f is shown in Fig. 121, curve I. It will be

* See Gray's *Absolute Measurements in Electricity and Magnetism*, pp. 251, 252 (1921).

seen from this curve that as the applied frequency increases from zero, L/L_1 increases from unity to a maximum of 6·16 at $f = 0·93$, then rapidly diminishes and becomes zero at a value of f slightly greater than unity (1·00676). Between this value of f and 4·281, L is negative, the greatest negative value being $4·547 L_1$ at $f = 1·08$. With further increase of f, L/L_1

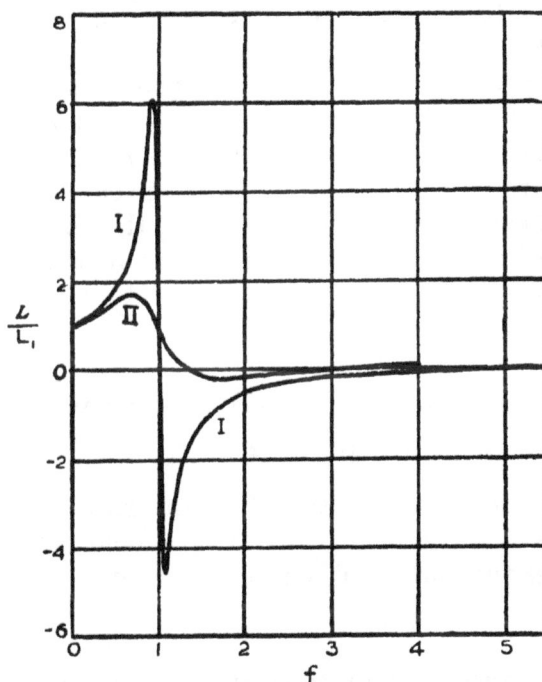

Fig. 121. Variation of Self-inductance of Transformer
with Frequency

becomes and remains positive, slowly approaching its final value 0·232.

The form of the curve in the neighbourhood of the resonance point ($f = 1$) depends greatly upon the value of the coefficient λ, which is proportional to the square of the resistance of the secondary circuit. If R_2 be regarded as the effective resistance of the secondary, as determined from the damping of its oscillation when the primary is on open circuit, the value of λ may vary considerably with the conditions of the experiment. Curve II in Fig. 121 shows the variation of L/L_1 when λ is 0·174, a value determined for the same induction coil in different circumstances. In this case L is negative between the values 1·338 and 3·222 of f, but the variation of L is now much smaller and more gradual near the resonance point. The greatest positive value of L/L_1 is 1·761 at $f = 0·7$, the greatest negative value 0·163 at $f = 1·7$. The effect of increasing the resistance of the secondary circuit is to diminish the range of variation of L, and also to diminish the range of frequencies over which L is negative.

If λ is increased beyond a certain point the L/L_1 curve fails to intersect the horizontal axis, and L can no longer have negative values. The limiting value of λ may be found by equating to zero the expression for L given in (127). The coupling k^2 being always less than unity, the conditions that the equation for f should have real positive roots are—

$$(\lambda + k^2)^2 \nless 4\lambda \ . \qquad . \qquad . \qquad . \qquad . \quad (129)$$

$$\text{and} \qquad \lambda + k^2 < 2. \qquad . \qquad . \qquad . \qquad . \quad (130)$$

The greatest value of λ that satisfies these conditions is

$$\lambda = 2 - k^2 - 2\sqrt{1 - k^2}. \qquad . \qquad . \qquad . \quad (131)$$

With $k^2 = 0·768$, this gives 0·2686 as the greatest value of λ which allows L to become zero. At this value of λ the L/L_1 curve touches the horizontal axis at the point—

$$f = 1/\sqrt{1 - k^2}.$$

If the resistance of the secondary circuit is so small that λ is negligible the expression (127) reduces to—

$$L = L_1\left(1 + \frac{k^2 f}{1 - f}\right) \qquad . \qquad . \qquad . \qquad . \quad (132)$$

which becomes infinite with change of sign at the resonance point, and changes sign again at $f = 1/(1 - k^2)$.

According to (127) the effect of an increase of the coupling is, in general, to increase the range of variation of L, and also to increase the range of frequencies over which L is negative. By (131) it also increases the maximum value of λ which allows L to change sign.

Application to Valve Experiment. In the valve experiment the rise of pitch sometimes heard in the telephone when the transformer is unshortcircuited is accounted for by the fact that the apparent self-inductance of the transformer primary is negative at certain frequencies. If the frequency with the transformer in circuit lies within the range over which L is negative the frequency will increase because of the diminution of self-inductance of the whole oscillatory circuit when the transformer is introduced.

In one experiment, in which the connections were as in Fig. 112, the pitch of the note was observed to fall slightly (L positive) when the induction coil was switched in if the capacity C_1 was less than about 0·19 mfd.; at this value the pitch was the same whether the coil was in circuit or not, indicating that L was zero. If C_1 was between 0·19 and 0·902 mfd. the pitch rose at unshortcircuiting (L negative), the rise of pitch at the latter value being about a semitone. The self-inductance of coil A in this experiment was 0·086 henry, so that a rise of pitch of a semitone indicates the addition to the circuit of a self-inductance of about – 0·01 henry. When C_1 was further slightly increased the pitch fell at the introduction of the coil by about the same interval, indicating a positive value for L, but it was noticed that the diminution of intensity of the sound at unshortcircuiting was considerably greater now than when L was negative. The current flowing in the oscillatory circuit was therefore considerably reduced in these circumstances, that is, just after L changed from a negative to a positive value.* At still greater values of C_1 a fall of pitch similar in amount was heard when the coil was thrown into circuit.

When the high-tension transformer was substituted for the induction coil the rise of pitch at unshortcircuiting could be made much greater. Thus, with $L_A = 0·086$ henry and $C_1 = 3·5$ mfd. the rise of pitch was about a minor sixth (see Fig. 114), indicating an apparent self-inductance of about – 0·052 henry. The value of the coefficient λ for the transformer in the circumstances of this experiment was about 0·52. It may be found from the period and decrement of the oscillation of the secondary circuit with the primary open (see Fig.

---

* In the circumstances of this experiment the core of the induction coil was in a state well below that of maximum inductance, so that a diminution of current was accompanied by a diminution of L_1.

40). With this value of λ it follows from condition (129) that the coupling of the transformer circuits was at least as great as 0·92.

Experiments on the effect of switching in the transformer may also be made by adjusting a variable condenser connected to the secondary terminals of the transformer, instead of varying C_1. If C_2 is increased, both λ and f increase, and the effect is somewhat similar to that of diminishing C_1. The primary

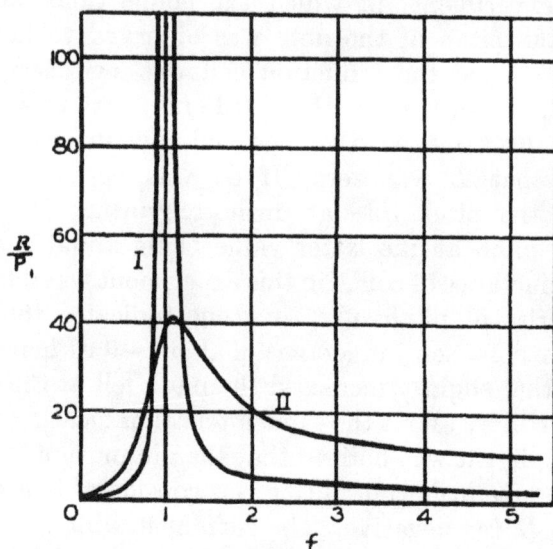

FIG. 122. VARIATION OF RESISTANCE OF TRANSFORMER WITH FREQUENCY

capacity being kept constant, if C_2 is small the apparent inductance of the transformer is negative, as shown by a rise of pitch at unshortcircuiting, if C_2 is large L is positive. At a certain value of the secondary capacity no change of pitch occurs when the transformer is short-circuited.

Variation of Apparent Resistance with Frequency. As to the apparent resistance of the primary coil as given by equation (128), the variation of the ratio R/R_1 with f is shown in Fig. 122 for the two numerical cases of Fig. 121, the value of R_1/L_1 being 149 in both cases. Curve I in Fig. 122 is the curve for the smaller secondary resistance ($\lambda = 0.005147$). In this case, the ratio R/R_1 is unity at very low frequency, 2·109 at very high frequency, and has a sharp maximum of 216·3 (not shown in the diagram) at a frequency slightly greater than the resonance value ($f = 1$). In curve II, calculated for the higher secondary resistance ($\lambda = 0 \cdot 174$), the variation is much more gradual, the maximum being about 41·6 at $f = 1 \cdot 1$, but the region of high apparent resistance extends to frequencies differing more widely from the resonance value. At $f = 2$, for example, the apparent resistance is over 20 times the resistance of the primary coil. The facts that the apparent resistance of the primary coil of a transformer increases as f approaches unity, and that there are in general, two values of f (one greater the other less than unity) at which R has the same value, are of importance in connection with the question, to be discussed later, of the maintenance of one or both of the oscillations by a valve.

Energy Dissipated in Transformer Circuits. If the time t be measured from a moment at which the primary current is zero, the expression for this current may be written—

$$i_1 = A \sin pt.$$

The energy dissipated in the primary circuit during one period T of the forced oscillation is—

$$\int_0^T R_1 i_1{}^2 dt = R_1 i_s{}^2 T,$$

i_s being the R.M.S. primary current.

From equations (122), (123) we find that the secondary current (here assumed to be uniformly distributed over the secondary winding, so that $L_{12} = L_{21} = M$) is—

$$i_2 = \frac{M p^2 C_2}{\{(1-f)^2 + \lambda f\}^{1/2}} A \sin (pt - \gamma),$$

where $\tan \gamma = \dfrac{R_2 C_2 p}{1 - f}$.

The energy dissipated in the secondary circuit during one period is therefore—

$$\int_0^T R_2 i_2{}^2 dt = \frac{T}{2} \frac{R_2 k^2 L_1 L_2 C_2{}^2 p^4 A^2}{(1 - f)^2 + \lambda f}$$

$$= \frac{R_2}{L_2} \cdot \frac{k^2 L_1 f^2}{(1 - f)^2 + \lambda f} i_s{}^2 T.$$

Consequently the total energy dissipated in one period is—

$$\int^T (R_1 i_1{}^2 + R_2 i_2{}^2)\, dt = R_1 \left\{ 1 + \frac{k^2 \dfrac{L_1}{R_1} \cdot \dfrac{R_2}{L_2} f^2}{(1-f)^2 + \lambda f} \right\} i_s{}^2 T$$

$$= R i_s{}^2 T. \qquad . \qquad . \qquad \text{by (128)}$$

The expression $R i_s{}^2$ therefore represents the mean rate of dissipation of energy in both circuits of the transformer during the forced oscillations. It also represents the mean rate at which work must be done on the system by the applied E.M.F. in order to maintain the oscillations. If R is very great, as it may be when the frequency is not far from the resonance value, it is clear that a valve, to which energy is supplied at a limited rate, may, for this reason, fail to maintain oscillations of this frequency and of appreciable amplitude in the transformer.

The Triode Oscillatory Circuit as a Coupled System. In the above discussion of apparent self-inductance and resistance we have regarded the transformer as executing forced oscillations under the action of an applied E.M.F. of frequency $n\ (= p/2\pi)$. When we deal with the whole oscillatory circuit of Fig. 112, however, of which the transformer forms a portion, we must regard n as a natural frequency, since it is usually determined mainly by the constants of the circuit.

We shall assume that the circuit possesses two natural frequencies of oscillation given, with sufficient accuracy, by the usual expression (13), page 6, which becomes in the present case—

$$8\pi^2 n^2 = \frac{1}{1-K^2} \left\{ \frac{1}{l_1 C_1} + \frac{1}{L_2 C_2} \right.$$

$$\left. \pm \sqrt{\left(\frac{1}{l_1 C_1} - \frac{1}{L_2 C_2}\right)^2 + 4K^2 \frac{1}{l_1 C_1} \frac{1}{L_2 C_2}} \right\}* \quad . \quad (133)$$

where l_1 is the self-inductance of the primary circuit, $L_1 + L_A$,

---

* In certain circumstances the frequency may, owing chiefly to the effects of grid current, differ widely from the value given by the inductances and capacities of the oscillatory circuit (see a paper by D. F. Martyn, *Phil. Mag.*, p. 922, November, 1927).

and K^2 is the coupling of the two circuits forming the whole system, i.e.

$$K^2 = L_{21}L_{12}/l_1L_2.$$

If L_2C_2/l_1C_1 be denoted by m, the ratios of the squares of the frequencies of these oscillations to the square of the

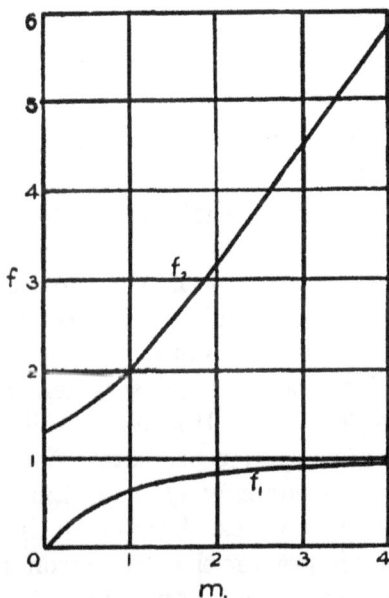

FIG. 123. VARIATION OF FREQUENCIES OF CIRCUIT WITH m.
COUPLING 0·26

The ordinate represents the square of the ratio of the lower or higher frequency of the system to the frequency of the secondary circuit when oscillating alone.

frequency of the secondary circuit when oscillating alone are given by—

$$f_1 = 4\pi^2 n_1^2 L_2 C_2 = \frac{1}{2(1 - K^2)}\left\{m + 1 - \sqrt{(m - 1)^2 + 4K^2m}\right\}$$
$$f_2 = 4\pi^2 n_2^2 L_2 C_2 = \frac{1}{2(1 - K^2)}\left\{m + 1 + \sqrt{(m - 1)^2 + 4K^2m}\right\}$$

. . . (134)

The manner in which f_1 and f_2 vary with m is illustrated by the two curves in Fig. 123, which represent values calculated from (134) for the particular case $K^2 = 0·26$, the coupling found for one of the arrangements used in the present experiments. The curves show that f_1 and f_2 continually increase as m increases from zero. At first f_1 increases rapidly, afterwards more slowly, gradually approaching unity, but only attaining this value when m becomes infinite. On the other hand, f_2 begins at the value $1 \cdot 3514$ $\left(\text{i.e. } \dfrac{1}{1 - K^2} \right)$ and increases to infinity as m indefinitely increases.

It appears, therefore, that the system (supposed to have resistances too small to affect the frequencies appreciably) never oscillates with a frequency lying between the values $1/2\pi\sqrt{L_2 C_2}$ (the resonance frequency of the transformer) and $1/2\pi\sqrt{L_2 C_2 (1 - K^2)}$, but that the frequency may have any value lying outside these limits.* Also the resonance frequency $(f = 1)$ only appears when m is very great, e.g. when the capacity C_1 is extremely small. The ratio of the squares of the two frequencies of the circuit, i.e. the ratio f_2/f_1, varies with the value of m, being smallest when $m = 1$, the ratio then being $(1 + K)/(1 - K)$.† It is evident also from the curves of Fig. 123 that if the system oscillates with a frequency less than the resonance frequency $(f < 1)$, this oscillation must be the slower of the two oscillations of the system; if the observed frequency is greater than the resonance value $(f > 1)$ this must be the more rapid component.

It is also clear that if the telephone note undergoes no change of pitch when the transformer is switched into the circuit of Fig. 112, this can only happen if the frequency before admission of the transformer is identical with that of the more rapid component afterwards, since L is zero only at values of f greater than unity (see Fig. 121). The condition $L = 0$, i.e. $1 - f + k^2 f = 0$, requires, in fact, that the frequency $1/2\pi\sqrt{L_1 C_1}$ should be equal to the greater of the two frequencies given by (133).

Effect of Variation of Primary Capacity. If, with the arrangement of Fig. 112 the capacity C_1 is gradually increased from a

---

* It may be noticed that the range $f = 1$ to $f = 1/(1 - K^2)$ is also that over which the apparent self-inductance of the *whole oscillatory circuit* (excluding the condenser C_1) is negative. This follows from equation (132) on substitution of K^2 for k^2.

† See equation (14), p. 7.

small value with the induction coil always in circuit, a single note is heard, high in pitch at first but falling continuously until a certain stage is reached at which a sudden fall of pitch occurs. After that the pitch falls gradually again with further increase of C_1.

Evidently, what happens in this experiment is that at first the upper note of the system is sounding, the pitch gradually descending along the upper curve of Fig. 123 (m diminishing) until at a certain point it suddenly changes to the lower curve

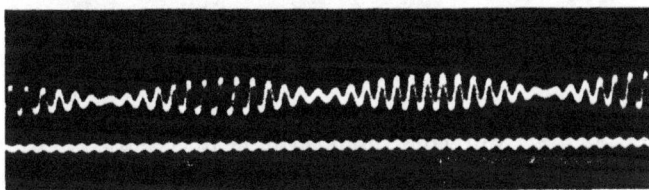

Fig. 124. Current Oscillogram Showing Simultaneous Existence of Both Component Oscillations

along which it continues to fall as m is further reduced. The experiment may also be made by first adjusting C_1 nearly to the critical value, when a slight change of the filament current produces the sudden change of pitch. A change in the filament current alters the current in the oscillatory circuit, and therefore also the inductances of the induction-coil circuits. Probably this effect is more marked in the secondary than in the primary, since the latter circuit contains also considerable air-core inductance. In this way a very gradual variation of the tuning of the circuits can be effected, and with this fine adjustment it is possible to have the two notes of the system sounding together in the telephone at a certain value of the filament current. Fig. 124 is a curve given by the current oscillograph O_1 (Fig. 112), Fig. 125 one shown by the electrostatic oscillograph O_2, when the two notes were sounding together. The coexistence of the two oscillations is clearly shown in these photographs by the "beats" which they produce.

In these experiments, however, the two notes sounded simultaneously only in a certain particular adjustment of the system, and it remains to be explained why the two notes are not *generally* produced together, why the lower note is heard when m is small (C_1 large), the higher note when m is large, and what determines the value of m at which the sudden change

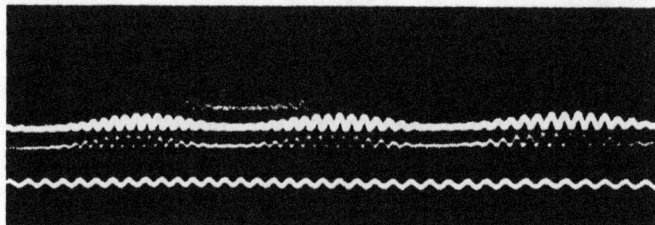

FIG. 125. SHOWING SIMULTANEOUS COMPONENT OSCILLATIONS IN SECONDARY CIRCUIT

of pitch occurs. For this purpose it is necessary to know the values of R, the apparent resistance of the transformer primary for various corresponding values of f_1 and f_2.

Apparent Resistances of the Transformer. Taking again the two numerical cases of Figs. 121 and 122, the values of R

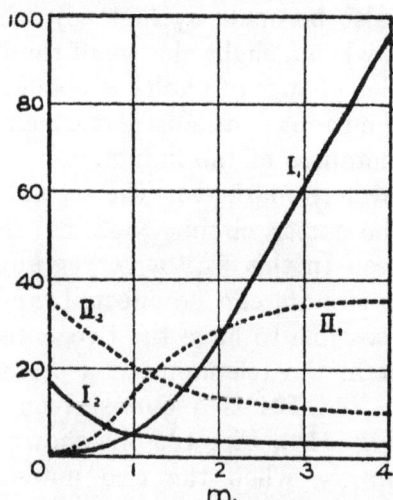

FIG. 126. RESISTANCE OF TRANSFORMER FOR THE TWO FREQUENCIES OF THE CIRCUIT

were calculated for the f_1 and f_2 corresponding to various values of m as shown in Fig. 123. The results are shown in Fig. 126, in which the full-line curves refer to the smaller secondary resistance ($\lambda = 0.005147$). In Fig. 126, curve I_1 shows values of R/R_1 for the oscillation of lower frequency (n_1) of the system, curve I_2 for the higher frequency (n_2). The curves extend over the range $m = 0$ to $m = 4$. It will be seen that when m is small the apparent resistance of the transformer primary is greater for the higher frequency than for the lower; when m is large R is greater for the lower frequency. The curves intersect at a value of m slightly greater than 1, and the values of R are varying rapidly at the point of intersection. Thus, at $m = 1.1$ the apparent resistance is nearly 40 per cent greater for the lower frequency component than for the higher; at $m = 0.9$ the apparent resistance is greater for the more rapid component in about the same ratio. At values of m more remote from unity the values of R for the two oscillations of the system differ much more. At $m = 1.5$, for example, R is 3.5 times as great for the slower oscillation as for the more rapid; at $m = 0.5$, R is over five times as great for the quicker component as for the slower. At the point of intersection the two equal values of R are each about $5.2 R_1$.

The broken line curves II_1, II_2 in Fig. 126 refer to the other numerical case ($\lambda = 0.174$). These curves present similar features; when m is small the apparent resistance is much greater for the quicker component, when m is large it is much smaller for this component. The point of intersection of the curves II_1, II_2, occurs at a value of m slightly greater than 1.2, the equal values of R being about $18.5 R_1$. It appears that the value of m at which the apparent resistance of the transformer primary is the same for the two oscillations of the system is rather greater than unity, its excess above this value increasing with the resistance of the secondary coil.

When we remember that the condition for maintenance of valve oscillations depends upon the resistance of the oscillatory circuit—the addition of a few ohms to this circuit is in some cases sufficient to stop the oscillations altogether—it is easy to understand from the curves of Fig. 126 why only one oscillation is usually maintained at a time, not both together, and why the slower component is the favoured one if m is small, the other if m is large.* Beginning with a small value of C_1 (m large) the oscillation at first is the higher note because the apparent resistance of the transformer for this oscillation (curve I_2, Fig. 126) is comparatively small, while for the lower note (curve I_1) the apparent resistance is very much greater. As C_1 is increased, the pitch gradually falling, the apparent resistance increases along the curve I_2 (Fig. 126), and is increasing rather rapidly when the point of intersection with I_1 (near $m = 1$) is reached. At this stage the apparent resistance is the same for the two oscillations, but a very slight further diminution of m causes the resistance to be appreciably greater for the rapid oscillation (curve I_2) than for the slower. Consequently, the oscillation, taking the form for which the resistance is smaller and the stability therefore greater, changes to the slower component, with sudden drop in pitch, the apparent resistance afterwards diminishing along the curve I_1 as m is further diminished.

It is clear that the sudden change of pitch must occur at or near the value of m for the point of intersection of the two curves of apparent resistance of the transformer, and that the simultaneous maintenance of the two oscillations can also only occur at the point of intersection, i.e. when the two notes are equally stable. Since the point of intersection of the curves occurs at a value of m not far from unity (see Fig. 126), and since the frequency ratio of a system of given coupling is least when $m = 1$, we may conclude that the two oscillations of the system can only be maintained simultaneously when the ratio of their frequencies is nearly at its minimum.

It may be remarked that similar effects can be obtained with other kinds of system, e.g. the singing arc with coupled circuits. The sudden change of pitch, and the simultaneous sounding of the two notes (with a very prominent difference-tone)

* It is well known that a valve fails to generate oscillations if the self-inductance in the primary circuit exceeds a certain value depending upon the other constants of the circuit. It might therefore be thought that the extinction of one of the component oscillations is due to the increase of L near the resonance point (see Fig. 121). This supposition would, however, not account for the facts. For if L is large and positive, f must be less than 1, and the frequency must consequently be that of the slower component (see Fig. 123). If the increased apparent self-inductance of the transformer primary were the cause of the extinction of one of the notes, this would always be the lower note.

at a value of m near unity, can be easily produced with this arrangement.*

It has been explained that the two notes can only be maintained simultaneously if the apparent resistance of the transformer is the same for the two frequencies of the system, that is at the intersection of the resistance curves of Fig. 126, but another condition is also necessary, viz. that the resistance at the intersection should not be too great. If R is too great at the intersection, no oscillations may be produced in appreciable intensity unless m is either considerably smaller or considerably greater than the value at the intersection. In one experiment, for example, in which the induction coil and the condenser C_1 were connected to the point X (Fig. 112) instead of to Y, a clear note was heard in the telephone with C_1 at 0·85 mfd., and another note a perfect fifth lower in pitch when C_1 was 1·56 mfd., but no note at all was heard with the induction coil in circuit if C_1 was between these two values. When C_1 was less than 0·85 mfd., the apparent resistance was on the right-hand and lower portion of a curve such as II_2 (Fig. 126), the note being the higher note of the system : when C_1 was above 1·56 mfd., the resistance was on the earlier portion of curve II_1, and the lower note was sounding.

From the above considerations and illustrations it appears that the principal phenomena associated with the maintenance of electrical oscillations in coupled circuits by a valve, viz. the production of one oscillation only in general, the sudden change of frequency, and the simultaneous maintenance of both oscillations in certain circumstances, can be adequately accounted for in terms of the "apparent resistance" of the transformer, and without reference to the characteristic properties of the valve. The valve is, in fact, here regarded merely as a device for transferring energy to the circuit at a limited rate and at a frequency determined by the constants of the circuit. There are, however, other phenomena, in particular the so-called "hysteresis" of valve oscillations, that is, the occurrence of the sudden change of frequency at rather different values of C_1 when this capacity is increasing and when it is diminishing, in the explanation of which the characteristics

* See *Phil. Mag.*, Jan., 1909, p. 41; Nov., 1909, p. 713; Oct., 1910, p. 660.

of the valve are involved.* Such effects were not prominent in the experiments described in the present chapter.

Transient Effects on Switching in Transformer. The explanation given in the previous paragraphs of the absence of one or other of the two component vibrations of the system applies only to the steady state some time after unshortcircuiting when the amplitude has become constant. In the period of transition immediately following the introduction of the transformer, however, while the system is accommodating itself to

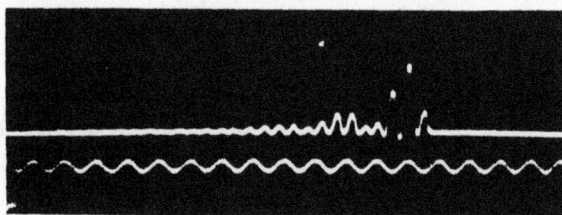

FIG. 127. SECONDARY POTENTIAL OF COIL WHEN SWITCHED
INTO CIRCUIT

its new conditions, there is evidence that the two components often exist together.

Fig. 127 is an example of the curves shown by the electrostatic oscillograph connected to the secondary terminals as in Fig. 112, when the key K was opened. In this experiment an induction coil was used in place of the transformer owing to its greater suitability for inductance measurements. The periodic variations of amplitude in the curve of Fig. 127 show clearly that two oscillations differing in frequency are present in the wave of secondary potential following the admission of the induction coil into the circuit. The wave form shows considerable variations, however, depending upon the phase of the previously existing oscillation at which the induction coil is switched in. If contact is broken at K at a moment when current is flowing into the condenser, a spark appears at the interrupter and the amplitude of the curve is small. The amplitude is greatest when contact is broken at a moment of maximum potential in C_1, and when consequently no spark

---

* For a discussion of these phenomena see a paper by B. van der Pol, Jr. *Phil. Mag.*, p. 700 (April, 1922).

appears at K. The curve shown in Fig. 127 was the curve of greatest amplitude among twenty photographs obtained with a certain combination of coils and condensers, all taken with the same current (indicated by the hot-wire ammeter) flowing in the condenser circuit before admission of the induction coil.

We will suppose, therefore, that the transformer is unshort-circuited at a moment when the condenser C_1 is at its maximum potential. At this instant there is no current or potential in the secondary circuit, and the conditions are very similar to those existing in the Tesla coil, which is excited by charging the condenser to a certain initial potential V_0 and allowing it to discharge through the primary coil.

We should, therefore, expect that when contact is broken at K (Fig. 112), two damped oscillations are set going in the circuit, the potential at the secondary terminals being given (see Chapter VII) by the expression—

$$V_2 = \frac{L_{21}C_1V_0}{\{(l_1C_1 - L_2C_2)^2 + 4K^2l_1C_1L_2C_2\}^{1/2}} \{e^{-k_1 t} \cos(2\pi n_1 t + e_1)$$
$$- e^{-k_2 t} \cos(2\pi n_2 t + e_2)\}, \qquad . \ (135)$$

where the damping coefficients k_1, k_2, are those given in equations (46) and (47), page 85, and e_1, e_2, are small angles which can be calculated from the constants of the circuits.*

In the present problem, however, there is another fact which must be taken into account. In the experiment of Fig. 127, the value of m ($= L_2C_2/l_1C_1$) was 1·507, and the apparent resistance of the coil was rather large, even for the more rapid component vibration (see Fig. 126). In the circumstances of the

---

* See Drude, *Ann. d. Physik*, 13, p. 539 (1904). The expressions for e_1, e_2, for a Tesla coil are—

$$\sin e_1 = \frac{2\beta + \dfrac{n_2 - n_1}{n_2}\sigma}{\dfrac{n_2 - n_1}{2\pi n_1 n_2} - 4\pi\beta n_1\sigma}$$

$$\sin e_2 = \frac{2\beta + \dfrac{n_2 - n_1}{n_1}\sigma}{\dfrac{n_2 - n_1}{2\pi n_1 n_2} - 4\pi\beta n_2\sigma}$$

where $\qquad \sigma = \dfrac{\theta_1 + \theta_2}{2} + \beta\dfrac{n_2 - n_1}{n_2 + n_1}$

experiment the valve was, in fact, unable to maintain either oscillation at a constant amplitude, as shown by the fact that the amplitude falls to zero after the transient in Fig. 127, though it considerably reduced the decrement of the more rapid component. It was necessary, therefore, to determine the value of k_2, the damping factor of the more rapid component, as modified by the action of the valve. This value was found from the primary current oscillogram, a curve, similar to that of Fig. 114, showing (after unshortcircuiting) a single damped oscillation* of frequency equal to that of the more rapid component of the system. From this curve the value of k_2 was

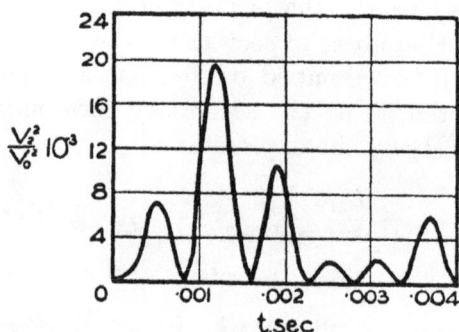

FIG. 128. SECONDARY POTENTIAL OF COIL AT ADMISSION INTO CIRCUIT (CALCULATED)

found to be 105, whereas the value calculated from the resistances of the circuits was 761. The difference represents the action of the valve in reducing the damping of the higher frequency oscillation. With this exception (the damping of the slower component being not appreciably affected by the valve), the constants of the circuits were all determined by the methods described in Chapter III. From these, when inserted in the expression (135), were calculated values of V_2 for various times t after the moment of admission of the induction coil into the circuit. The values of V_2 being squared, for comparison with

---

* It may seem remarkable that the primary current curve at unshortcircuiting shows no evidence of double periodicity such as that plainly shown in the curve of secondary potential of Fig. 127. A calculation of the initial amplitudes of the two components of the primary current, as given in equation (98), p. 179, showed however that that of the slow vibration was only one quarter of that of the rapid one, while its decrement was considerably the greater of the two. The slower component was, therefore, not much in evidence in the primary circuit.

the oscillograph curve (Fig. 127), the calculated results are shown in Fig. 128, in which the horizontal axis represents the time t in seconds, the ordinate the ratio of the square of V_2 to that of V_0, the maximum potential of the primary condenser in the oscillations preceding the opening of the key K. A comparison of the observed and calculated curves of Figs. 127 and 128 shows that there is good agreement between them, in regard to both the time intervals between the successive zeroes and the relative heights of the successive peaks in the curves.

The maximum potential indicated in Fig. 128 is 140·3 V_0. Now, the smallest R.M.S. primary current which was observed to give at unshortcircuiting a 1-cm. spark between two metal spheres of 2 cm. diameter connected with the secondary terminals was 1·57 amp. Also, the frequency of the preceding oscillations was 753·6, and the capacity of the primary condensor 2·17 mfd. Consequently, the value of V_0 in this experiment was—

$$V_0 = \frac{1 \cdot 57\sqrt{2} \cdot 10^6}{2\pi \cdot 753 \cdot 6 \cdot 2 \cdot 17}$$

$$= 216 \text{ volts.}$$

The calculated maximum secondary potential is therefore—

$$V_{2m} = 140 \cdot 3 \times 216$$

$$= 30,300 \text{ volts,}$$

which does not fall far short of the generally accepted value of 31,000 volts required to spark across a 1-cm. gap of this kind.

It may be remarked that if the same primary condenser C_1 were charged by a battery to potential V_0 and then discharged through the primary of the coil, the maximum secondary potential in this case would be 96·4 V_0. The effect of the valve in the experiment of Fig. 127 is to raise this to 140·3 V_0, and it does this simply by reducing the damping factor of one of the two vibrations of the coupled system.

From the close agreement between the observed and calculated results in this experiment it may be concluded that the application here made of the theory of the Tesla coil, modified by substitution of the observed damping coefficient of the higher frequency vibration of the system for the value calculated from the resistances, accounts satisfactorily for the abnormal potential effects observed when the primary of a transformer or coil is switched into a triode oscillatory circuit.

The nature of the transient effect here discussed, which occurs when a transformer is switched into a triode circuit, is very different from that illustrated in Fig. 53, page 109, which shows the effect of connecting a transformer to a source of ordinary A.C. current. In the experiment of Fig. 53, the transient includes one oscillation, and it dies away as the alternating current is becoming established. In the problem discussed in the present chapter, the transient consists of two oscillations, one of which subsides to zero, and the other may or may not afterwards persist as a regular periodic flow of current in the circuit.

APPENDIX

ELECTRON DIFFRACTION
BY NITRO-CELLULOSE FILMS

AT the end of Chapter VI it is stated that diffraction patterns similar to those obtained with celluloid films are also observed when the pencil of cathode rays is transmitted through sufficiently thin films of pure nitro-cellulose. One of the photographs obtained with this substance is reproduced in Fig. 129.

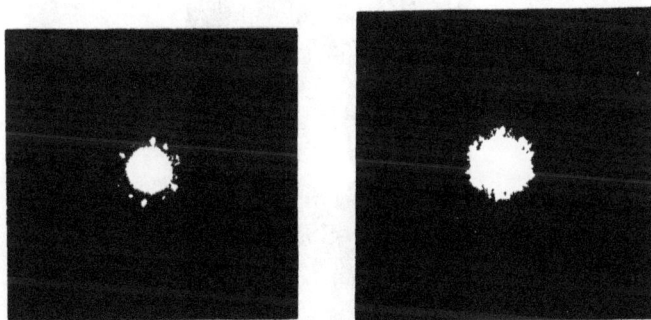

FIGS. 129 AND 130. NITRO-CELLULOSE

It shows the first ring of the pattern with twelve maxima equally spaced round its circumference, and the pattern is evidently of the same type as that of Fig. 84, page 155.

In addition to the six or twelve principal maxima on the first ring, these patterns usually show a number of finer maxima, and in some cases the secondary maxima show evidence of regularity of arrangement in relation to the principal maxima. In Fig. 130, for example, taken with another film of nitro-cellulose, the secondary maxima are arranged in six groups on the ring. The regularity of arrangement of the secondary maxima is, however, shown more clearly by films of celluloid (probably because they are usually thinner), as, for example, in the photograph reproduced in Fig. 131.

The photographs reproduced in Figs. 129, 130, 131, were taken with a camera similar to that described in Chapter VI

241

(Fig. 83, page 153), but larger, the distance from film to plate being about 36 cm. The peak potential in Fig. 129 was 80 kV, in Fig. 130, 70 kV, and in Fig. 131, 80 kV.

As to the meaning of the secondary maxima, if it is correct to assume, as suggested in Chapter VI, that the six or twelve principal maxima arise from scattering by centres near the two sides of the film, it seems natural to suppose that the secondary maxima are due to the internal portions of the film. If this view is correct, it appears that the regularity of arrangement of the scattering centres exists throughout the whole thickness

FIG. 131. SHOWING ARRANGEMENT OF SECONDARY MAXIMA IN CELLULOID DIFFRACTION PATTERN

of the film, the planes of centres in different layers being, however, not quite parallel to each other.

Another feature of the pattern in Fig. 131 is that the numbers of secondary maxima are not the same in all the groups. In two of the diametrically opposite pairs of groups there are five maxima, in the third pair there are ten.* These numbers might suggest that there is some connection between the secondary maxima and the arrangement of the oxygen and hydrogen atoms in the cellulose molecule, but there are other details, regarding the arrangement of the maxima within each group, which require further investigation.

* These details are shown better in the original negative from which Fig. 131 was prepared.

INDEX

PRINTED IN GREAT BRITAIN AT THE PITMAN PRESS, BATH
C2—(5729)

AN ABRIDGED LIST OF

TECHNICAL BOOKS

PUBLISHED BY

Sir Isaac Pitman & Sons, Ltd.

Parker Street, Kingsway, London, W.C.2

The prices given apply only to Great Britain

A complete Catalogue giving full details of the following books will be sent post free on application

---

CONTENTS

C2—4

ALL PRICES ARE NET

THE ARTISTIC CRAFT SERIES

s. d.

BOOKBINDING AND THE CARE OF BOOKS. By Douglas Cockerell.
Fourth Edition 7 6
DRESS DESIGN. By Talbot Hughes 12 6
EMBROIDERY AND TAPESTRY WEAVING. By Mrs. A. H. Christie.
Fourth Edition 10 6
HAND-LOOM WEAVING. By Luther Hooper 10 6
HERALDRY. By Sir W. H. St. John Hope, Litt.D., D.C L. . 12 6
SILVERWORK AND JEWELLERY. By H. Wilson. Second Edition 10 6
STAINED GLASS WORK. By C. W. Whall 10 6
WOOD-BLOCK PRINTING. By F. Morley Fletcher . . . 8 6
WOODCARVING DESIGN AND WORKMANSHIP. By George Jack.
Second Edition 8 6
WRITING AND ILLUMINATING AND LETTERING. By Edward
Johnston. Sixteenth Edition 8 6

ART AND CRAFT WORK, ETC.

BLOCK-CUTTING AND PRINT-MAKING BY HAND. By Margaret
Dobson, A.R.E. 12 6
CABINET-MAKING, THE ART AND CRAFT OF. By D. Denning . 5 0
CELLULOSE LACQUERS. By S. Smith, O.B.E., Ph.D. . . 7 6
HANDICRAFTS, HOME DECORATIVE. By Mrs. F. Jefferson-
Graham 25 0
LACQUER WORK. By G. Koizumi 15 0
LEATHER CRAFT, ARTISTIC. By Herbert Turner . . . 5 0
LEATHER WORK: STAMPED, MOULDED, CUT, CUIR-BOUILLI,
SEWN, ETC. By Charles G. Leland. Third Edition . . 5 0
LETTERING, DECORATIVE WRITING AND ARRANGEMENT OF. By
Prof. A. Erdmann and A. A. Braun. Second Edition. . 10 6
LETTERING AND DESIGN, EXAMPLES OF. By J. Littlejohns, R.B.A. 4 0
LETTERING, PLAIN AND ORNAMENTAL. By Edwin G. Fooks . 3 6
MANUSCRIPT AND INSCRIPTION LETTERS. By Edward Johnston. 7 6
MANUSCRIPT WRITING AND LETTERING. By an Educational
Expert 6 0
METAL WORK. By Charles G. Leland. Second Edition. . 5 0
ORNAMENTAL HOMECRAFTS. By Idalia B. Littlejohns . . 10 6
PLYWOOD AND GLUE, MANUFACTURE AND USE OF. By B. C.
Boulton, B.Sc. 7 6
POTTERY, HANDCRAFT. By H. and D. Wren. . . . 12 6
STAINED GLASS, THE ART AND CRAFT OF. By E. W. Twining . 42 0
STENCIL-CRAFT. By Henry Cadness, F.S.A.M. . . . 10 6
WEAVING FOR BEGINNERS. By Luther Hooper . . . 5 0
WEAVING WITH SMALL APPLIANCES—
THE WEAVING BOARD. By Luther Hooper . . . 7 6
TABLE LOOM WEAVING. By Luther Hooper . . . 7 6
TABLET WEAVING. By Luther Hooper 7 6
WOOD CARVING. By Charles G. Leland. Fifth Edition . . 5 0
WOODCARVING, HANDICRAFT OF. By James Jackson . . 4 0

TEXTILE MANUFACTURE, ETC.

	s.	d.
ARTIFICIAL SILK. By Dr. V. Hottenroth. Translated from the German by Dr. E. Fyleman, B.Sc.	30	0
ARTIFICIAL SILK. By Dr. O. Faust. Translated by Dr. E. Fyleman	10	6
ARTIFICIAL SILK AND ITS MANUFACTURE. By Joseph Foltzer. Translated into English by T. Woodhouse. 4th Ed.	21	0
ARTIFICIAL SILK OR RAYON, ITS MANUFACTURE AND USES. By T. Woodhouse, F.T.I. Second Edition	7	6
ARTIFICIAL SILK OR RAYON, THE PREPARATION AND WEAVING OF. By T. Woodhouse, F.T.I.	10	6
BLEACHING, DYEING, PRINTING, AND FINISHING FOR THE MANCHESTER TRADE. Bv J. W. McMyn, F.C.S., and J. W. Bardsley. Second Edition	6	0
COLOUR IN WOVEN DESIGN. By Roberts Beaumont, M.Sc., M.I.Mech.E. Second Edition, Revised and Enlarged.	21	0
COTTON SPINNER'S POCKET BOOK, THE. By James F. Innes. Third Edition	3	6
COTTON SPINNING COURSE, A FIRST YEAR. By H. A. J. Duncan, A.T.I.	5	0
COTTON WORLD, THE. Compiled and Edited by J. A. Todd, M.A., B.L.	5	0
DRESS, BLOUSE, AND COSTUME CLOTHS. DESIGN AND FABRIC MANUFACTURE. By Roberts Beaumont, M.Sc., M.I.Mech.E., and Walter G. Hill	42	0
FLAX AND JUTE, SPINNING, WEAVING, AND FINISHING OF. By T. Woodhouse, F.T.I.	10	6
FLAX CULTURE AND PREPARATION. By F. Bradbury. 2nd Ed.	10	6
FUR. By MAX BACHRACH, B.C.S.	21	0
FURS AND FURRIERY. By Cyril J. Rosenberg	30	0
HOSIERY MANUFACTURE. By Prof. W. Davis, M.A. 2nd Ed.	5	0
KNITTED FABRICS, CALCULATIONS AND COSTINGS FOR. By Professor William Davis, M.A.	10	6
LOOM, THEORY AND ELECTRICAL DRIVE OF THE. By R. H. Wilmot, M.Sc., A.M.I.E.E., Assoc.A.I.E.E.	8	6
MEN'S CLOTHING, ORGANIZATION, MANAGEMENT, AND TECHNOLOGY IN THE MANUFACTURE OF. By M. E. Popkin.	25	0
PATTERN CONSTRUCTION, THE SCIENCE OF. For Garment Makers. By B. W. Poole	45	0
TEXTILE CALCULATIONS. By J. H. Whitwam, B.Sc. (Lond.)	25	0
TEXTILE EDUCATOR, PITMAN'S. Edited by L. J. Mills, *Fellow of the Textile Institute.* In three volumes	63	0
TEXTILES FOR SALESMEN. By E. Ostick, M.A., L.C.P.	5	0
TEXTILES, INTRODUCTION TO. By A. E. Lewis, A.M.C.T., A.T.I.	3	6
TOWELS AND TOWELLING, THE DESIGN AND MANUFACTURE OF. By T. Woodhouse, F.T.I., and A. Brand, A.T.I.	12	6
WEAVING AND MANUFACTURING, HANDBOOK OF. By H. Greenwood, A.T.I.	5	0
WOOLLEN YARN PRODUCTION. By T. Lawson	3	6
WOOL SUBSTITUTES, By Roberts Beaumont, M.Sc., M.I.Mech.E.	10	6

Textile Manufacture, etc.—contd.

	s.	d.
WOOL, THE MARKETING OF. By A. F. DuPlessis, M.A.	12	6
WORSTED OPEN DRAWING. By S. Kershaw, F.T.I.	5	0
YARNS AND FABRICS, THE TESTING OF. By H. P. Curtis. 2nd Ed.	5	0

DRAUGHTSMANSHIP

	s.	d.
BLUE PRINT READING. By J. Brahdy, B.Sc., C.E.	10	6
DRAWING AND DESIGNING. By Charles G. Leland, M.A. Fourth Edition	3	6
DRAWING OFFICE PRACTICE. By H. Pilkington Ward, M.Sc., A.M.Inst.C.E.	7	6
ENGINEER DRAUGHTSMEN'S WORK. By A Practical Draughtsman	2	6
ENGINEERING DESIGN, EXAMPLES IN. By G. W. Bird, B.Sc. Second Edition	6	0
ENGINEERING HAND SKETCHING AND SCALE DRAWING. By Thos. Jackson, M.I.Mech.E., and Percy Bentley, A.M.I.Mech.E.	3	0
ENGINEERING WORKSHOP DRAWING. By A. C. Parkinson, B.Sc. Second Edition	4	0
MACHINE DRAWING, A PREPARATORY COURSE TO. By P. W. Scott	2	0
PLAN COPYING IN BLACK LINES. By B. J. Hall, M.I.Mech.E.	2	6

PHYSICS, CHEMISTRY, ETC.

	s.	d.
ARTIFICIAL RESINS. By J. Scheiber, Ph.D. Translated by E. Fyleman, B.Sc., Ph.D., F.I.C.	30	0
BIOLOGY, INTRODUCTION TO PRACTICAL. By N. Walker.	5	0
BOTANY, TEST PAPERS IN. By E. Drabble, D.Sc.	2	0
CHEMICAL ENGINEERING, AN INTRODUCTION TO. By A. F. Allen, B.Sc. (Hons.), F.C S., LL.B.	10	6
CHEMISTRY, A FIRST BOOK OF. By A. Coulthard, B.Sc. (Hons.), Ph.D., F.I.C.	3	0
CHEMISTRY, DEFINITIONS AND FORMULAE FOR STUDENTS. By W. G. Carey, F.I.C.	–	6
CHEMISTRY, TEST PAPERS IN. By E. J. Holmyard, M.A.	2	0
With Points Essential to Answers	3	0
CHEMISTRY, HIGHER TEST PAPERS IN. By the same Author. 1. Inorganic. 2. Organic. Each	3	0
DISPENSING FOR PHARMACEUTICAL STUDENTS. By J. W. Cooper and F. J. Dyer, Second Edition	7	6
ELECTRICITY AND MAGNETISM, FIRST BOOK OF. By W. Perren Maycock, M.I.E.E. Fourth Edition.	6	0
ENGINEERING PRINCIPLES, ELEMENTARY. By G. E. Hall, B.Sc.	2	6
ENGINEERING SCIENCE, A PRIMER OF. By Ewart S. Andrews, B.Sc. (Eng.).		
Part I. FIRST STEPS IN APPLIED MECHANICS.	2	6
II. FIRST STEPS IN HEAT AND HEAT ENGINES	2	0
LATIN FOR PHARMACEUTICAL STUDENTS. By J. W. Cooper and A. C. Mclaren	6	0

CONSTRUCTIONAL ENGINEERING

s. d.

REINFORCED CONCRETE, DETAIL DESIGN IN. By Ewart S.
Andrews, B.Sc. (Eng.) 6 0
REINFORCED CONCRETE. By W. Noble Twelvetrees, M.I.M.E.,
A.M.I.E.E. 21 0
REINFORCED CONCRETE MEMBERS, SIMPLIFIED METHODS OF
CALCULATING. By W. Noble Twelvetrees. Second Edition. 5 0
SPECIFICATIONS FOR BUILDING WORKS. By W. L. Evershed, F.S.I. 5 0
STRUCTURES, THE THEORY OF. By H. W. Coultas, M.Sc.,
A.M.I.Struct.E., A.I.Mech.E. 15 0

CIVIL ENGINEERING, BUILDING, ETC.

AUDEL'S MASONS' AND BUILDERS' GUIDES. In four volumes
Each 7 6
1. BRICKWORK, BRICK-LAYING, BONDING, DESIGNS
2. BRICK FOUNDATIONS, ARCHES, TILE SETTING, ESTIMATING
3. CONCRETE MIXING, PLACING FORMS, REINFORCED
STUCCO
4. PLASTERING, STONE MASONRY, STEEL CONSTRUCTION,
BLUE PRINTS
AUDEL'S PLUMBERS' AND STEAM FITTERS' GUIDES. Practical
Handbooks in four volumes . . . . . Each 7 6
1. MATHEMATICS, PHYSICS, MATERIALS, TOOLS, LEAD-
WORK
2. WATER SUPPLY, DRAINAGE, ROUGH WORK, TESTS
3. PIPE FITTING, HEATING, VENTILATION, GAS, STEAM
4. SHEET METAL WORK, SMITHING, BRAZING, MOTORS
BRICKWORK, CONCRETE, AND MASONRY. Edited by T. Corkhill,
M.I.Struct.E. In eight volumes . . . . Each 6 0
" THE BUILDER " SERIES—
ARCHITECTURAL HYGIENE; OR, SANITARY SCIENCE AS
APPLIED TO BUILDINGS. By Sir Banister Fletcher,
F.R.I.B.A., F.S.I., and H. Phillips Fletcher, F.R.I.B.A.,
F.S.I. Fifth Edition, Revised 10 6
CARPENTRY AND JOINERY. By Sir Banister Fletcher,
F.R.I.B.A., F.S.I., etc., and H. Phillips Fletcher,
F.R.I.B.A., F.S.I., etc. Fifth Edition, Revised . . 10 6
QUANTITIES AND QUANTITY TAKING. By W. E. Davis.
Seventh Edition, Revised by P. T. Walters, F.S.I., F.I.Arb. 6 0
BUILDING, DEFINITIONS AND FORMULAE FOR STUDENTS. By T.
Corkhill, F.B.I.C.C., M.I.Struct.E. . . . . . – 6
BUILDING EDUCATOR, PITMAN'S. Edited by R. Greenhalgh,
A.I.Struct.E. In three volumes 63 0
FIELD MANUAL OF SURVEY METHODS AND OPERATIONS. By A.
Lovat Higgins, B.Sc., A.R.C.S., A.M.I.C.E. . . . 21 0
HYDRAULICS. By E. H. Lewitt, B.Sc. (Lond.), M.I.Ae.E.,
A.M.I.M.E. Fourth Edition 10 6
HYDRAULICS FOR ENGINEERS. By Robert W. Angus, B.A.Sc. 12 6

Civil Engineering, Building, etc.—contd.

	s.	d.
JOINERY & CARPENTRY. Edited by R. Greenhalgh,A.I.Struct.E. In six volumes . . . . Each	6	0
MECHANICS OF BUILDING. By Arthur D. Turner, A.C.G.I., A.M.I.C.E.	5	0
PAINTING AND DECORATING. Edited by C. H. Eaton, F.I.B.D. In six volumes . . . . Each	7	6
PLUMBING AND GASFITTING. Edited by Percy Manser, R.P., A.R.San.I. In seven volumes . . . . Each	6	0
SURVEYING, TUTORIAL LAND AND MINE. By Thomas Bryson	10	6
WATER MAINS, LAY-OUT OF SMALL. By H. H. Hellins, M.Inst.C.E.	7	6
WATERWORKS FOR URBAN AND RURAL DISTRICTS. By H. C. Adams, M.Inst.C.E., M.I.M.E., F.S.I. Second Edition. .	15	0

MECHANICAL ENGINEERING

	s.	d.
AUDEL'S ENGINEERS' AND MECHANICS' GUIDES. In eight volumes. Vols. 1–7 . . . . . . Each	7	6
Vol. 8	15	0
CONDENSING PLANT. By R. J. Kaula, M.I.E.E., and I. V. Robinson, Wh.Sc., A.M.Inst.C.E.	30	0
DEFINITIONS AND FORMULAE FOR STUDENTS—APPLIED MECHANICS. By E. H. Lewitt, B.Sc., A.M.I.Mech.E.. . .	–	6
DEFINITIONS AND FORMULAE FOR STUDENTS—HEAT ENGINES. By A. Rimmer, B.Eng. Second Edition. . . .	–	6
DIESEL ENGINES : MARINE, LOCOMOTIVE, AND STATIONARY. By David Louis Jones, *Instructor, Diesel Engine Department, U.S. Navy Submarine Department*	21	0
ENGINEERING EDUCATOR, PITMAN'S. Edited by W. J. Kearton, M.Eng., A.M.I.Mech.E., A.M.Inst.N.A. In three volumes	63	0
ESTIMATING FOR MECHANICAL ENGINEERS. By L. E. Bunnett, A.M.I.P.E.	10	6
FRICTION CLUTCHES. By R. Waring-Brown, A.M.I.A.E., F.R.S.A., M.I.P.E.	5	0
FUEL ECONOMY IN STEAM PLANTS. By A. Grounds, B.Sc., F.I.C., F.Inst.P.	5	0
FUEL OILS AND THEIR APPLICATIONS. By H. V. Mitchell, F.C.S. Second Edition, Revised by A. Grounds, B.Sc., A.I.C.	5	0
MECHANICAL ENGINEERING DETAIL TABLES. By P. Ross .	7	6
MECHANICAL ENGINEER'S POCKET BOOK, WHITTAKER'S. Third Edition, entirely rewritten and edited by W. E. Dommett, A.F.Ae.S., A.M.I.A.E.	12	6
MECHANICS' AND DRAUGHTSMEN'S POCKET BOOK. By W. E. Dommett, Wh.Ex., A.M.I.A.E.	2	6
MECHANICS FOR ENGINEERING STUDENTS. By G. W. Bird, B.Sc., A.M.I.Mech.E., A.M.I.E.E. Second Edition . .	5	0
MECHANICS OF MATERIALS, EXPERIMENTAL. By H. Carrington, M.Sc. (Tech.), D.Sc., M.Inst.Met., A.M.I.Mech.E.,A.F.R.Æ.S.	3	6

Mechanical Engineering—contd.

AERONAUTICS, ETC.

Aeronautics, etc.—contd.

s. d.

CIVILIAN AIRCRAFT, REGISTER OF. By W. O. Manning and
R. L. Preston 3 6
FLYING AS A CAREER. By Major Oliver Stewart, M.C., A.F.C. 3 6
GLIDING AND MOTORLESS FLIGHT. By C. F. Carr and L.
Howard-Flanders, A.F.R.Æ.S., M.I.Æ.E., A.M.I.Mech.E. . 7 6
LEARNING TO FLY. By F. A. Swoffer, M.B.E. With a Foreword
by the late Sir Sefton Brancker, K.C.B., A.F.C. 2nd Ed. . 7 6
PARACHUTES FOR AIRMEN. By Charles Dixon . . . 7 6
PILOT's " A " LICENCE Compiled by John F. Leeming, *Royal
Aero Club Observer for Pilot's Certificates.* Fourth Edition . 3 6

MARINE ENGINEERING

MARINE ENGINEERING, DEFINITIONS AND FORMULAE FOR
STUDENTS. By E. Wood, B.Sc. . . . . . – 6
MARINE SCREW PROPELLERS, DETAIL DESIGN OF. By Douglas
H. Jackson, M.I.Mar.E., A.M.I.N.A. 6 0

OPTICS AND PHOTOGRAPHY

AMATEUR CINEMATOGRAPHY. By Capt. O. Wheeler, F.R.P.S. . 6 0
APPLIED OPTICS, AN INTRODUCTION TO. Volume I. By L. C.
Martin, D.Sc., D.I.C., A.R.C.S. 21 0
BROMOIL AND TRANSFER. By L. G. Gabriel 7 6
CAMERA LENSES. By A. W. Lockett 2 6
COLOUR PHOTOGRAPHY. By Capt. O. Wheeler, F.R.P.S.. . 12 6
COMMERCIAL PHOTOGRAPHY. By D. Charles 5 0
COMPLETE PRESS PHOTOGRAPHER, THE. By Bell R. Bell. . 6 0
LENS WORK FOR AMATEURS. By H. Orford. Fifth Edition,
Revised by A. Lockett 3 6
PHOTOGRAPHIC CHEMICALS AND CHEMISTRY. By J. South-
worth and T. L. J. Bentley 3 6
PHOTOGRAPHIC PRINTING. By R. R. Rawkins . . . 3 6
PHOTOGRAPHY AS A BUSINESS. By A. G. Willis . . . 5 0
PHOTOGRAPHY THEORY AND PRACTICE. By E. P. Clerc. Edited
by G. E. Brown 35 0
RETOUCHING AND FINISHING FOR PHOTOGRAPHERS. By J. S.
Adamson 4 0
STUDIO PORTRAIT LIGHTING. By H. Lambert, F.R.P.S. . . 15 0

ASTRONOMY

ASTRONOMY, PICTORIAL. By G. F. Chambers, F.R.A.S.. . 2 6
ASTRONOMY FOR EVERYBODY. By Professor Simon Newcomb,
LL.D. With an Introduction by Sir Robert Ball . . 5 0
GREAT ASTRONOMERS. By Sir Robert Ball, D.Sc., LL.D., F.R.S. 5 0
HIGH HEAVENS, IN THE. By Sir Robert Ball . . . 5 0
STARRY REALMS, IN. By Sir Robert Ball, D.Sc., LL.D., F.R.S. 5 0

MOTOR ENGINEERING

	s.	d.
AUTOMOBILE AND AIRCRAFT ENGINES By A. W. Judge, A.R.C.S., A.M.I.A.E. Second Edition	42	0
CARBURETTOR HANDBOOK, THE. By E. W. Knott, A.M.I.A.E..	10	6
GAS AND OIL ENGINE OPERATION. By J. Okill, M.I.A.E.. .	5	0
GAS, OIL, AND PETROL ENGINES. By A. Garrard, Wh.Ex. .	6	0
MAGNETO AND ELECTRIC IGNITION. By W. Hibbert, A.M.I.E.E. Third Edition	3	6
MOTOR-CYCLIST'S LIBRARY, THE. Each volume in this series deals with a particular type of motor-cycle from the point of view of the owner-driver . . . . . Each	2	0

> A.J.S., THE BOOK OF THE. By W. C. Haycraft.
> ARIEL, THE BOOK OF THE. By G. S. Davison.
> B.S.A., THE BOOK OF THE. By "Waysider."
> DOUGLAS, THE BOOK OF THE. By E. W. Knott.
> IMPERIAL, BOOK OF THE NEW. By F. J. Camm.
> MATCHLESS, THE BOOK OF THE. By W. C. Haycraft.
> NORTON, THE BOOK OF THE. By W. C. Haycraft
> P. AND M., THE BOOK OF THE. By W. C. Haycraft.
> RALEIGH HANDBOOK, THE. By "Mentor."
> ROYAL ENFIELD, THE BOOK OF THE. By R. E. Ryder.
> RUDGE, THE BOOK OF THE. By L. H. Cade.
> TRIUMPH, THE BOOK OF THE. By E. T. Brown.
> VILLIERS ENGINE, BOOK OF THE. By C. Grange.

MOTORISTS' LIBRARY, THE. Each volume in this series deals with a particular make of motor-car from the point of view of the owner-driver. The functions of the various parts of the car are described in non-technical language, and driving repairs, legal aspects, insurance, touring, equipment, etc., all receive attention.

	s.	d.
AUSTIN, THE BOOK OF THE. By B. Garbutt. Third Edition, Revised by E. H. Row	3	6
MORGAN, THE BOOK OF THE. By G. T. Walton . .	2	6
SINGER JUNIOR, BOOK OF THE. By G. S. Davison. .	2	6
MOTORIST'S ELECTRICAL GUIDE, THE. By A. H. Avery, A.M.I.E.E.	3	6
CARAVANNING AND CAMPING. By A. H. M. Ward, M.A.	2	6

ELECTRICAL ENGINEERING, ETC.

	s.	d.
ACCUMULATOR CHARGING, MAINTENANCE, AND REPAIR. By W. S. Ibbetson. Second Edition	3	6
ALTERNATING CURRENT BRIDGE METHODS. By B. Hague, D.Sc. Second Edition	15	0

Electrical Engineering, etc.—contd.

Electrical Engineering, etc.—contd.

Electrical Engineering, etc.—contd.

Telegraphy, Telephony, and Wireless—contd.

MATHEMATICS AND CALCULATIONS FOR ENGINEERS

Mathematics for Engineers—contd.

	s.	d.
MATHEMATICS, INDUSTRIAL (PRELIMINARY), By G. W. Stringfellow	2	0
With Answers	2	6
MEASURING AND MANURING LAND, AND THATCHER'S WORK, TABLES FOR. By J. Cullyer. Twentieth Impression . .	3	0
MECHANICAL TABLES. By J. Foden	2	0
MECHANICAL ENGINEERING DETAIL TABLES. By John P. Ross	7	6
METALWORKER'S PRACTICAL CALCULATOR, THE. By J. Matheson	2	0
METRIC CONVERSION TABLES. By W. E. Dommett, A.M.I.A.E.	1	0
METRIC LENGTHS TO FEET AND INCHES, TABLE FOR THE CONVERSION OF. Compiled by Redvers Elder. On paper. .	1	0
MINING MATHEMATICS (PRELIMINARY). By George W. Stringfellow	1	6
With Answers	2	0
QUANTITIES AND QUANTITY TAKING. By W. E. Davis. Seventh Edition, Revised by P. T. Walters, F.S.I., F.I.Arb. . .	6	0
SCIENCE AND MATHEMATICAL TABLES. By W. F. F. Shearcroft, B.Sc., A.I.C., and Denham Larrett, M.A. . . .	1	0
SLIDE RULE, THE. By C. N. Pickworth, Wh.Sc. Seventeenth Edition, Revised	3	6
SLIDE RULE : ITS OPERATIONS ; AND DIGIT RULES, THE. By A. Lovat Higgins, A.M.Inst.C.E.	–	6
STEEL'S TABLES. Compiled by Joseph Steel	3	6
TELEGRAPHY AND TELEPHONY, ARITHMETIC OF. By T. E. Herbert, M.I.E.E., and R. G. de Wardt . . .	5	0
TEXTILE CALCULATIONS. By J. H. Whitwam, B.Sc. . .	25	0
TRIGONOMETRY FOR ENGINEERS, A PRIMER OF. By W. G. Dunkley, B.Sc. (Hons.)	5	0
TRIGONOMETRY FOR NAVIGATING OFFICERS. By W. Percy Winter, B.Sc. (Hons.), Lond.	10	6
TRIGONOMETRY, PRACTICAL. By Henry Adams, M.I.C.E., M.I.M.E., F.S.I. Third Edition, Revised and Enlarged .	5	0
VENTILATION, PUMPING, AND HAULAGE, MATHEMATICS OF. By F. Birks	5	0
WORKSHOP ARITHMETIC, FIRST STEPS IN. By H. P. Green .	1	0

MISCELLANEOUS TECHNICAL BOOKS

	s.	d.
BOOT AND SHOE MANUFACTURE. By F. Plucknett . . .	35	0
BREWING AND MALTING. By J. Ross Mackenzie, F.C.S., F.R.M.S. Second Edition	8	6
BUILDER'S BUSINESS MANAGEMENT. By J. H. Bennetts, A.I.O.B.	10	6
CERAMIC INDUSTRIES POCKET BOOK. By A. B. Searle . .	8	6
CINEMA ORGAN, THE. By Reginald Foort, F.R.C.O. . .	2	6
ELECTRICAL HOUSECRAFT. By R. W. Kennedy . . .	2	6
ENGINEERING ECONOMICS. By T. H. Burnham, B.Sc. (Hons.), B.Com., A.M.I.Mech.E. Second Edition	10	6

Miscellaneous Technical Books—contd.

	s.	d.
ENGINEERING INQUIRIES, DATA FOR. By J..C. Connan, B.Sc., A.M.I.E.E., O.B.E.	12	6
ESTIMATING. By T. H. Hargrave. Second Edition	7	6
FARADAY, MICHAEL, AND SOME OF HIS CONTEMPORARIES. By Prof. William Cramp, D.Sc., M.I.E.E.	2	6
FURNITURE STYLES. By. H. E. Binstead. Second Edition	10	6
GLUE AND GELATINE. By P. I. Smith.	8	6
GRAMOPHONE HANDBOOK. By W. S. Rogers	2	6
HAIRDRESSING, THE ART AND CRAFT OF. Edited by G. A. Foan.	60	0
HIKER AND CAMPER, THE COMPLETE. By C. F. Carr	2	6
HOUSE DECORATIONS AND REPAIRS. By W. Prebble. Second Edition	1	0
MOTOR BOATING. By F. H. Snoxell	2	6
MUSIC ENGRAVING AND PRINTING. By Wm. Gamble, F.R.P.S.	21	0
PAPER TESTING AND CHEMISTRY FOR PRINTERS. By Gordon A. Jahans, B.A.	12	6
PETROLEUM. By Albert Lidgett. Third Edition	5	0
PRINTING. By H. A. Maddox	5	0
REFRACTORIES FOR FURNACES, CRUCIBLES, ETC. By A. B. Searle	5	0
REFRIGERATION, MECHANICAL. By Hal Williams, M.I.Mech.E., M.I.E.E., M.I.Struct.E. Third Edition	20	0
SEED TESTING. By J. Stewart Remington	10	6
STONES, PRECIOUS AND SEMI-PRECIOUS. By Michael Weinstein. Second Edition	7	6
STORES ACCOUNTS AND STORES CONTROL. By J. H. Burton. Second Edition	5	0
TALKING PICTURES. By Bernard Brown, B.Sc. (Eng.)	12	6
TEACHING METHODS FOR TECHNICAL TEACHERS. By J. H. Currie, M.A., B.Sc., A.M.I.Mech.E. .	2	6

PITMAN'S TECHNICAL PRIMERS

	s.	d.
Each in foolscap 8vo, cloth, about 120 pp., illustrated	2	6

In each book of the series the fundamental principles of some subdivision of technology are treated in a practical manner, providing the student with a handy survey of the particular branch of technology with which he is concerned.

ABRASIVE MATERIALS. By A. B. Searle.

A.C. PROTECTIVE SYSTEMS AND GEARS. By J. Henderson, B.Sc., M.C., and C. W. Marshall, B.Sc., M.I.E.E.

BELTS FOR POWER TRANSMISSION. By W. G. Dunkley, B.Sc.

BOILER INSPECTION AND MAINTENANCE. By R. Clayton.

CAPSTAN AND AUTOMATIC LATHES. By Philip Gates.

CENTRAL STATIONS, MODERN. By C. W. Marshall, B.Sc., A.M.I.E.E.

COAL CUTTING MACHINERY, LONGWALL. By G. F. F. Eagar, M.I.Min.E.

CONTINUOUS CURRENT ARMATURE WINDING. By F. M. Denton, A.C.G.I., A.Amer.I.E.E.

CONTINUOUS CURRENT MACHINES, THE TESTING OF. By Charles F. Smith, D.Sc., M.I.E.E., A.M.I.C.E.

Pitman's Technical Primers—contd. Each 2s. 6d.

COTTON SPINNING MACHINERY AND ITS USES. By Wm. Scott Taggart, M.I.Mech.E.

DIESEL ENGINE, THE. By A. Orton, A.M.I.Mech.E.

DROP FORGING AND DROP STAMPING. By H. Hayes.

ELECTRIC CABLES. By F. W. Main, A.M.I.E.E.

ELECTRIC CRANES AND HAULING MACHINES. By F. E. Chilton, A.M.I.E.E.

ELECTRIC FURNACE, THE. By Frank J. Moffett, B.A., M.I.E.E.

ELECTRIC MOTORS, SMALL. By E. T. Painton, B.Sc., A.M.I.E.E.

ELECTRICAL INSULATION. By W. S. Flight, A.M.I.E.E.

ELECTRICAL TRANSMISSION OF ENERGY. By W. M. Thornton, O.B.E., D.Sc., M.I.E.E.

ELECTRICITY IN AGRICULTURE. By A. H. Allen, M.I.E.E.

ELECTRICITY IN STEEL WORKS. By Wm. McFarlane, B.Sc.

ELECTRIFICATION OF RAILWAYS, THE. By H. F. Trewman, M.A.

ELECTRO-DEPOSITION OF COPPER, THE. And its Industrial Applications. By Claude W. Denny, A.M.I.E.E.

EXPLOSIVES, MANUFACTURE AND USES OF. By R. C. Farmer, O.B.E., D.Sc., Ph.D.

FILTRATION. By T. R. Wollaston, M.I.Mech.E.

FOUNDRYWORK. By Ben Shaw and James Edgar.

GRINDING MACHINES AND THEIR USES. By Thos. R. Shaw, M.I.Mech.E.

HYDRO-ELECTRIC DEVELOPMENT. By J. W. Meares, F.R.A.S., M.Inst.C.E., M.I.E.E., M.Am.I.E.E.

ILLUMINATING ENGINEERING, THE ELEMENTS OF. By A. P. Trotter, M.I.E.E.

INDUSTRIAL AND POWER ALCOHOL. By R. C. Farmer, O.B.E., D.Sc., Ph.D., F.I.C.

INDUSTRIAL ELECTRIC HEATING. By J. W. Beauchamp, M.I.E.E.

INDUSTRIAL MOTOR CONTROL. By A. T. Dover, M.I.E.E.

INDUSTRIAL NITROGEN. By P. H. S. Kempton, B.Sc. (Hons.), A.R.C.Sc.

KINEMATOGRAPH STUDIO TECHNIQUE. By L. C. Macbean.

LUBRICANTS AND LUBRICATION. By J. H. Hyde.

MECHANICAL HANDLING OF GOODS, THE. By C. H. Woodfield, M.I.Mech.E.

MECHANICAL STOKING. By D. Brownlie, B.Sc., A.M.I.M.E. (Double volume, price 5s. net.)

METALLURGY OF IRON AND STEEL. Based on Notes by Sir Robert Hadfield.

MUNICIPAL ENGINEERING. By H. Percy Boulnois, M.Inst.C.E., F.R.San.Inst., F.Inst.S.E.

OILS, PIGMENTS, PAINTS, AND VARNISHES. By R. H. Truelove.

PATTERNMAKING. By Ben Shaw and James Edgar.

PETROL CARS AND LORRIES. By F. Heap.

PHOTOGRAPHIC TECHNIQUE. By L. J. Hibbert, F.R.P.S.

PNEUMATIC CONVEYING. By E. G. Phillips, M.I.E.E., A.M.I.Mech.E.

Pitman's Technical Primers—contd. Each 2s. 6d.

POWER FACTOR CORRECTION. By A. E. Clayton, B.Sc. (Eng.) Lond., A.K.C., A.M.I.E.E.

RADIOACTIVITY AND RADIOACTIVE SUBSTANCES. By J. Chadwick, M.Sc., Ph.D.

RAILWAY SIGNALLING: AUTOMATIC. By F. Raynar Wilson.

RAILWAY SIGNALLING: MECHANICAL. By F. Raynar Wilson.

SEWERS AND SEWERAGE. By H. Gilbert Whyatt, M.I.C.E.

SPARKING PLUGS. By A. P. Young and H. Warren.

STEAM ENGINE VALVES AND VALVE GEARS. By E. L. Ahrons, M.I.Mech.E., M.I.Loco.E.

STEAM LOCOMOTIVE, THE. By E. L. Ahrons, M.I.Mech.E., M.I.Loco.E.

STEAM LOCOMOTIVE CONSTRUCTION AND MAINTENANCE. By E. L. Ahrons, M.I.Mech.E., M.I.Loco.E.

STEELWORK, STRUCTURAL. By Wm. H. Black.

STREETS, ROADS, AND PAVEMENTS. By H. Gilbert Whyatt, M.Inst.C.E., M.R.San.I.

SWITCHBOARDS, HIGH TENSION. By Henry E. Poole, B.Sc. (Hons.), Lond., A.C.G.I., A.M.I.E.E.

SWITCHGEAR, HIGH TENSION. By Henry E. Poole, B.Sc.(Hons.), A.C.G.I., A.M.I.E.E.

SWITCHING AND SWITCHGEAR. By Henry E. Poole, B.Sc.(Hons.), A.C.G.I., A.M.I.E.E.

TELEPHONES, AUTOMATIC. By F. A. Ellson, B.Sc., A.M.I.E.E. (Double volume, price 5s.)

TIDAL POWER. By A. M. A. Struben, O.B.E., A.M.Inst.C.E.

TOOL AND MACHINE SETTING. For Milling, Drilling, Tapping, Boring, Grinding, and Press Work. By Philip Gates.

TOWN GAS MANUFACTURE. By Ralph Staley, M.C.

TRACTION MOTOR CONTROL. By A. T. Dover, M.I.E.E.

TRANSFORMERS AND ALTERNATING CURRENT MACHINES, THE TESTING OF. By Charles F. Smith, D.Sc.. A.M.Inst.C.E.

TRANSFORMERS, HIGH VOLTAGE POWER. By Wm. T. Taylor, M.Inst.C.E., M.I.E.E.

TRANSFORMERS, SMALL SINGLE-PHASE. By Edgar T. Painton, B.Sc. Eng. (Hons.) Lond., A.M.I.E.E.

WATER POWER ENGINEERING. By F. F. Fergusson, C.E., F.G.S., F.R.G.S.

WIRELESS TELEGRAPHY, CONTINUOUS WAVE. By B. E. G. Mittell, A.M.I.E.E.

WIRELESS TELEGRAPHY, DIRECTIVE. Direction and Position Finding, etc. By L. H. Walter, M.A. (Cantab.), A.M.I.E.E.

X-RAYS, INDUSTRIAL APPLICATION OF. By P. H. S. Kempton, B.Sc. (Hons.), A.R.C.S.

COMMON COMMODITIES AND INDUSTRIES

Each book in crown 8vo, illustrated. **3s.** net.

In each of the handbooks in this series a particular product or industry is treated by an expert writer and practical man of business. Beginning with the life history of the plant, or other natural product, he follows its development until it becomes a commercial commodity, and so on through the various phases of its sale in the market and its purchase by the consumer.

Asbestos. (SUMMERS.)

Bookbinding Craft and Industry. (HARRISON.)

Books—From the MS. to the Bookseller. (YOUNG.)

Boot and Shoe Industry, The. (HARDING.)

Bread and Bread Baking. (STEWART.)

Brushmaker, The. (KIDDIER.)

Butter and Cheese. (TISDALE and JONES.)

Button Industry, The. (JONES.)

Carpets. (BRINTON.)

Clays and Clay Products. (SEARLE.)

Clocks and Watches. (OVERTON.)

Clothing Industry, The. (POOLE.)

Cloths and the Cloth Trade. (HUNTER.)

Coal. (WILSON.)

Coal Tar. (WARNES.)

Coffee—From Grower to Consumer. (KEABLE.)

Cold Storage and Ice Making. (SPRINGETT.)

Concrete and Reinforced Concrete. (TWELVETREES.)

Copper—From the Ore to the Metal. (PICARD.)

Cordage and Cordage Hemp and Fibres. (WOODHOUSE and KILGOUR.)

Corn Trade, The British. (BARKER.)

Cotton. (PEAKE.)

Cotton Spinning. (WADE.)

Drugs in Commerce. (HUMPHREY.)

Dyes. (HALL.)

Electricity. (NEALE.)

Engraving. (LASCELLES.)

Explosives, Modern. (LEVY.)

Fertilizers. (CAVE.)

Fishing Industry, The. (GIBBS)

Furniture. (BINSTEAD.)

Furs and the Fur Trade. (SACHS)

Gas and Gas Making. (WEBBER.)

Glass and Glass Making. (MARSON)

Gloves and the Glove Trade. (ELLIS.)

Gold. (WHITE.)

Gums and Resins. (PARRY.)

Ink. (MITCHELL.)

Iron and Steel. (HOOD.)

Ironfounding. (WHITELEY.)

Jute Industry The. (WOODHOUSE and KILGOUR.)

Knitted Fabrics. (CHAMBERLAIN and QUILTER.)

Lead, including Lead Pigments. (SMYTHE.)

Leather. (ADCOCK.)

Linen. (MOORE.)

Locks and Lock Making. (BUTTER.)

Match Industry, The. (DIXON.)

Meat Industry The. (WOOD.)

Oils. (MITCHELL.)

Paints and Varnishes. (JENNINGS.)

Paper. (MADDOX.)

Perfumery, The Raw Materials of. (PARRY.)

Photography. (GAMBLE.)

Platinum Metals, The. (SMITH.)

Pottery. (NOKE and PLANT.)

Common Commodities and Industries—contd.

Rice. (DOUGLAS.)

Rubber. (STEVENS and STEVENS)

Salt. (CALVERT.)

Silk. (HOOPER.)

Soap. (SIMMONS.)

Sponges. (CRESSWELL.)

Starch and Starch Products. (AUDEN.)

Stones and Quarries. (HOWE.)

Sugar. (MARTINEAU.) (Revised by EASTICK.)

Sulphur and Allied Products. (AUDEN.)

Tea. (IBBETSON.)

Telegraphy, Telephony, and Wireless. (POOLE.)

Textile Bleaching. (STEVEN.)

Timber. (BULLOCK.)

Tin and the Tin Industry. (MUNDEY.)

Tobacco. (TANNER.) (Revised by DREW.)

Weaving. (CRANKSHAW.)

Wheat and Its Products. (MILLAR.)

Wine and the Wine Trade. (SIMON.)

Wool. (HUNTER.)

Worsted Industry, The. (DUMVILLE and KERSHAW.)

Zinc and Its Alloys. (LONES.)

PITMAN'S SHORTHAND
INVALUABLE TO ALL BUSINESS AND PROFESSIONAL MEN

The following Catalogues will be sent post free on application—

SCIENTIFIC AND TECHNICAL

EDUCATIONAL, COMMERCIAL, SHORTHAND

FOREIGN LANGUAGES, AND ART

PRINTED IN GREAT BRITAIN AT THE PITMAN PRESS, BATH
(3129w)

DEFINITIONS AND FORMULAE FOR STUDENTS

This series of booklets is intended to provide engineering students with all necessary definitions and formulae in a convenient form.

ELECTRICAL

By PHILIP KEMP, M.Sc., M.I.E.E., *Assoc.A.I.E.E., Head of the Electrical Engineering Department of the Regent Street Polytechnic.*

HEAT ENGINES

By ARNOLD RIMMER, B.Eng., *Head of the Mechanical Engineering Department, Derby Technical College.* Second Edition.

APPLIED MECHANICS

By E. H. LEWITT, B.Sc., A.M.I.Mech.E.

PRACTICAL MATHEMATICS

By LOUIS TOFT, M.Sc., *Head of the Mathematical Department of the Royal Technical College, Salford.*

CHEMISTRY

By W. GORDON CAREY, F.I.C.

BUILDING

By T. CORKHILL, F.B.I.C.C., M.I.Struct.E., M.Coll.H.

AERONAUTICS

By JOHN D. FRIER, A.R.C.Sc., D.I.C., F.R.Ae.S.

COAL MINING

By M. D. WILLIAMS, F.G.S.

MARINE ENGINEERING

By E. WOOD, B.Sc.

ELECTRICAL INSTALLATION WORK

By F. PEAKE SEXTON, A.R.C.S., A.M.I.E.E.

LIGHT AND SOUND

By P. K. BOWES, M.A., B.Sc.

Each in pocket size, about 32 pp. **6d.** net.

Sir Isaac Pitman & Sons, Ltd., Parker Street, Kingsway, W.C.2

PITMAN'S
TECHNICAL
DICTIONARY

OF

ENGINEERING *and* INDUSTRIAL SCIENCE

IN SEVEN LANGUAGES

ENGLISH, FRENCH, SPANISH, ITALIAN, PORTUGUESE, RUSSIAN, AND GERMAN

WITH AN ADDITIONAL VOLUME CONTAINING A COMPLETE KEY INDEX IN EACH OF THE SEVEN LANGUAGES

Edited by

ERNEST SLATER, M.I.E.E., M.I.Mech.E.

In Collaboration with Leading Authorities

THE Dictionary is arranged upon the basis of the English version. This means that against every English term will be found the equivalents in the six other languages, together with such annotations as may be necessary to show the exact use of the term in any or all of the languages.

" There is not the slightest doubt that this Dictionary will be the essential and standard book of reference in its sphere. It has been needed for years."—*Electrical Industries.*

" The work should be of the greatest value to all who have to deal with specifications, patents, catalogues, etc., for use in foreign trade." —*Bankers' Magazine.*

" The work covers extremely well the ground it sets out to cover, and the inclusion of the Portuguese equivalents will be of real value to those who have occasion to make technical translations for Portugal, Brazil, or Portuguese East Africa."—*Nature.*

Complete in five volumes. Crown 4to, buckram gilt, £8 8s. net.

SIR ISAAC PITMAN & SONS, LTD., PARKER STREET, KINGSWAY, W.C.2

www.ingramcontent.com/pod-product-compliance
Lightning Source LLC
Chambersburg PA
CBHW021425180326
41458CB00001B/140